ASPECTS OF COMBINATORICS

A WIDE-RANGING INTRODUCTION

Aspects of combinatorics

VICTOR BRYANT

Lecturer, Department of Pure Mathematics, University of Sheffield

Published by the Press Syndicate of the University of Cambridge
The Pitt Building, Trumpington Street, Cambridge CB2 1RP
40 West 20th Street, New York, NY 10011-4211, USA
10 Stamford Road, Oakleigh, Melbourne 3166, Australia

© Cambridge University Press 1992

First published 1992
Reprinted 1995

A catalogue record for this book is available from the British Library

Library of Congress cataloguing-in-publication data
Bryant, Victor.
Aspects of combinatorics : a wide-ranging introduction / Victor Bryant.
p. cm.
Includes bibliographical references and index.
ISBN 0 521 41974 3 (hc). – ISBN 0 521 42997 8 (pb)
1. Combinatorial analysis. I. Title.
QA164.B82   1992
511'.6 – dc20   92-6827 CIP

ISBN 0 521 41974 3   hardback
ISBN 0 521 42997 8   paperback

Transferred to digital printing 2000

KW

Contents

	Introduction	vii
1	The binomial coefficients	1
2	How many trees?	13
3	The marriage theorem	25
4	Three basic principles	36
5	Latin squares	50
6	The first theorem of graph theory	71
7	Edge-colourings	81
8	Harems and tournaments	91
9	Minimax theorems	98
10	Recurrence	109
11	Vertex-colourings	127
12	Rook polynomials	138
13	Planar graphs	151
14	Map-colourings	166
15	Designs and codes	179
16	Ramsey theory	201
	Hints for exercises	224
	Answers to exercises	249
	Bibliography	262
	Index	263

Introduction

By its formal dictionary definition 'combinatorics' is that branch of mathematics which deals with combinations, and those combinations can be of sets, graphs, matrices, traffic routes, people, etc. etc. Indeed, such an elementary theme is bound to have many applications and to cut across many other branches of mathematics, and that is one of the fascinations of the subject. It is also bound to have many different aspects, some of which are presented here.

Over a decade ago, when courses on this subject were first entering the undergraduate curriculum, combinatorics was often looked upon disparagingly as an easy option. But the popularity of such courses amongst undergraduates is due largely to the fact that at least the *questions* make sense, even though the *answers* are as difficult as in any other course. Another advantage to the student is that the wide range of topics means that even if one chapter is a complete haze the next one offers a fresh start. The advantage to the lecturer of this wide selection of ideas is that he or she can include many miscellaneous topics discovered when reading the latest books and journals.

Of course there are dangers that such a *pot-pourri* will become just a rag-bag of trivia, but as you collect together material for a course you begin to see how deep and inter-related some of the ideas are. Some of these inter-relationships only become clear after teaching the material for many years: there is real pleasure in marking a piece of work where a student is the first to notice that the 'doctor's waiting-room' example from chapter 1 can be used in solving the 'hostess problem' in chapter 12, and in discovering that a theorem from tournaments can be proved much more neatly by a simple application of a marriage theorem derived earlier.

My own selection of *Aspects*... is, of course, a very personal one and it includes a large amount of graph theory and transversal theory. The subjects are not grouped together but visited again and again throughout the text: I have found that this is the most entertaining way to present the material and that a pause before seeing another application of some theorem enables greater understanding of it. Although there are no formal pre-requisites for studying this material, a lot of it requires a certain mathematical maturity and is unlikely to be suitable for students before at least their second year in higher education. There is more than enough material here for a full year's course and the lecturer may choose to save some of the topics for a more advanced course or merely to omit some of the harder results and/or proofs.

I would like to place on record my thanks to Hazel Perfect who (together with the late Leon Mirsky) wrote some pioneering papers in transversal theory and who kindly and caringly involved me in their world when I was a newly qualified lecturer. My thanks also go to all the authors whose books and papers have coloured my choice of topics and the way in which I present them. I am also grateful to Peter Brooksbank, a Sheffield graduate now doing his own research, for reading the typescript and making some valuable comments. But most of all my wholehearted thanks go to the many other Sheffield undergraduates and postgraduates who, over the years, have attended my courses in combinatorics and who have shared my enthusiasm for the subject. Teaching them has been a very great pleasure and privilege.

Victor Bryant
1992

1
The binomial coefficients

The idea of choosing a number of items is going to occur many times in any course on combinatorics and so an appropriate first chapter is on selections and on the number of ways those selections can be made. In this chapter there are no theorems, simply a range of miscellaneous ideas presented as examples. Although the ideas are elementary and although many of them will be familiar, the chances are that most readers will find some unfamiliar aspect amongst them.

Given n objects let $\binom{n}{k}$ denote the number of different selections of k objects from the n: the order in which the k objects are chosen is irrelevant.

Example *There are 16 chapters in this book: how many different selections of two chapters are there? In general, in how many ways can two be chosen from n?*

Solution

$$\binom{16}{2} = \frac{16 \times 15}{2} = 120.$$

There are several ways of seeing this. For example you could list all pairs of chapters, with 16 choices for the first of the pair and 15 remaining choices for the other member of the pair:

1, 2
1, 3
1, 4
$\vdots$
1, 16
2, 1
2, 3
2, 4
$\vdots$
16, 14
16, 15

There are clearly 16×15 pairs in this list but note that any two chapters appear in the list as two different pairs. For example 7 and 11 will appear once as '7, 11' and then again as '11, 7'. Since the order is irrelevant to us we only want to count half of this list, making $\binom{16}{2}$ equal to 120 as above.

Another way of seeing that $\binom{16}{2} = 120$ is by again listing the pairs but avoiding a repeated pair by only writing down those (like '7, 11') where the first member of the pair is the lower:

$$\left.\begin{matrix}1,2\\1,3\\\vdots\\1,16\end{matrix}\right\}15$$

$$\left.\begin{matrix}2,3\\\vdots\\2,16\end{matrix}\right\}14$$

In this list each possible pair occurs exactly once and so we must simply count the number of pairs in the list to give

$$\binom{16}{2} = 15 + 14 + 13 + \cdots + 2 + 1 = \tfrac{1}{2} \times 15 \times 16$$

$$\left.\begin{matrix}3,4\\\vdots\\3,16\end{matrix}\right\}13$$

as before.

$$\vdots$$

$$\left.\begin{matrix}14,15\\14,16\end{matrix}\right\}2$$

$$15,16\}\ 1$$

Clearly both these arguments extend to choosing two from n to give

$$\binom{n}{2} = \tfrac{1}{2}n(n-1). \qquad \square$$

Example *In how many ways can three of the 16 chapters be chosen? Show that in general*

$$\binom{n}{3} = \binom{n-1}{2} + \binom{n-2}{2} + \binom{n-3}{2} + \cdots + \binom{3}{2} + \binom{2}{2}.$$

Solution Following the first of the methods in the previous solution gives a total list of $16 \times 15 \times 14$ triples of chapters; 16 choices for the first of the triple, 15 remaining choices for the second and 14 remaining choices for the third. But any particular set of three chapters will be in that long list six times (for example the set of three chapters 7, 11 and 15 will occur in the list as '7, 11, 15', '7, 15, 11', '11, 7, 15', '11, 15, 7', '15, 7, 11' and '15, 11, 7'). So the number of different selections of three chapters from the 16 is

$$\binom{16}{3} = \frac{16 \times 15 \times 14}{6} = 560.$$

The second method in the previous solution was to list the required selections with the lowest first to avoid repeated pairs. In a similar way we could now list the triples

The binomial coefficients

with each triple in increasing order: to count the number of triples in this list we can use the result of the previous example:

$$\left.\begin{array}{c} 1, 2, 3 \\ 1, 2, 4 \\ \vdots \\ 1, 15, 16 \end{array}\right\} \quad \binom{15}{2} \text{ of the triples begin with a 1,}$$

$$\left.\begin{array}{c} 2, 3, 4 \\ \vdots \\ 2, 15, 16 \end{array}\right\} \quad \binom{14}{2} \text{ of the triples begin with a 2,}$$

$$\left.\begin{array}{c} 3, 4, 5 \\ \vdots \\ 3, 15, 16 \end{array}\right\} \quad \binom{13}{2} \text{ of the triples begin with a 3,}$$

$$\vdots \qquad \vdots$$

$$\left.\begin{array}{c} 13, 14, 15 \\ 13, 14, 15 \\ 13, 15, 16 \end{array}\right\} \quad \binom{3}{2} \text{ of the triples begin with a 13,}$$

$$14, 15, 16 \} \quad \binom{2}{2} \text{ of the triples begin with a 14.}$$

Hence

$$\binom{16}{3} = \binom{15}{2} + \binom{14}{2} + \binom{13}{2} + \cdots + \binom{3}{2} + \binom{2}{2} = 560.$$

It is left to the reader to show that a similar argument choosing three items from n would yield the relationship

$$\binom{n}{3} = \binom{n-1}{2} + \binom{n-2}{2} + \binom{n-3}{2} + \cdots + \binom{3}{2} + \binom{2}{2}. \qquad \square$$

That gentle introduction leads to the idea that if k items are chosen from n, where the order of the choices matters, then the number of selections is

$$\underbrace{n \times (n-1) \times (n-2) \times \cdots}_{k \text{ numbers}}$$

But if the order in which the k are chosen does not matter then each set of k items will occur in that grand total

$$k \times (k-1) \times (k-2) \times \cdots \times 2 \times 1$$

times. This is abbreviated to $k!$ (called k *factorial*). So the number of different selections of k items from n is given by

$$\binom{n}{k} = \frac{n(n-1)(n-2)\cdots(n-k+1)}{k(k-1)(k-2)\cdots 1} = \frac{n!}{k!(n-k)!}$$

Of course we have assumed that n and k are positive integers with $k \leqslant n$. With the convention that $0! = 1$ the above formula still makes sense if either k or n is zero. If $k > n$ or $k < 0$ then we take $\binom{n}{k}$ to be 0.

Example *Let k and n be integers with $1 \leqslant k \leqslant n$. Use a selection argument to show that*

$$\binom{n}{k} = \binom{n-1}{k-1} + \binom{n-2}{k-1} + \binom{n-3}{k-1} + \cdots + \binom{k}{k-1} + \binom{k-1}{k-1}.$$

Solution The left-hand expression is simply the number of ways of choosing k numbers from $\{1, 2, \ldots, n\}$. How many of those collections have 1 as the lowest? How many have 2 as the lowest? The solution, which generalises the idea used in the previous examples, is now left to the reader. □

Example *How many x^k do you get in the expansion of $(1 + x)^n$?*

Solution
$$(1 + x)^n = \underbrace{(1 + x)(1 + x)(1 + x) \cdots (1 + x)}_{n \text{ lots}}.$$

We could prove this result by induction on n but instead we'll use a selection argument. Multiplying out $(1 + x)^n$ consists of adding up all the terms obtained by multiplying together one entry from the first bracket (namely either 1 or x) times one item from the second, times one from the third, etc. To get an x^k you must therefore have chosen the x from precisely k of the n brackets. So the total number of x^k in the expansion equals the number of ways of choosing k brackets from the n, namely $\binom{n}{k}$. □

It follows from that example that

$$(1 + x)^n = \binom{n}{0} + \binom{n}{1}x + \binom{n}{2}x^2 + \cdots + \binom{n}{k}x^k + \cdots + \binom{n}{n}x^n.$$

This is the well-known *binomial expansion* and so the numbers $\binom{n}{k}$ are often referred to as the *binomial coefficients*.

Example *Show that*

$$\binom{n}{0} + \binom{n}{1} + \binom{n}{2} + \cdots + \binom{n}{n} = 2^n.$$

The binomial coefficients

Solution The easiest way to derive this result is to put x equal to 1 in the binomial expansion above. But an alternative method by selection arguments is to count the number of subsets of a set of n items. There are $\binom{n}{0}$ no-element subsets (i.e. just the empty set), $\binom{n}{1}$ one-element subsets, $\binom{n}{2}$ two-element subsets, and so on. Hence the left-hand expression in the example represents the total number of subsets (of all sizes) of a set of n items. Is there a quicker way of counting those subsets? To form a subset you must simply decide for each of the n items whether it is in the subset or not. This is a two-way choice ('in' or 'out') and is made independently for each of the n items. So the total number of ways of choosing a subset is 2^n and the above result follows. □

Example *Imagine that this picture represents a rectangular grid of roads and that you want to walk along the roads from A to B using as short a route as possible. How many different such routes are there? Generalise the result to any size grid.*

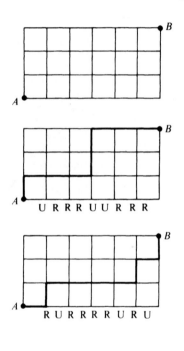

Solution Two typical routes from A to B are illustrated. Any shortest route from A to B must consist of nine moves towards B: in other words each route must consist of nine moves, any three of which must be 'up' (U) and the rest of which must be 'right' (R). For example, the two illustrated routes can be described by the sequences of 'ups' and 'rights' stated. So the number of acceptable routes from A to B equals the number of ways of choosing which three of the nine moves must be 'up'. There are $\binom{9}{3} = 84$ such choices.

A similar argument shows that the number of shortest routes from A to B in the grid illustrated on the right (with n moves necessary, any k of which must be 'up') is $\binom{n}{k}$. □

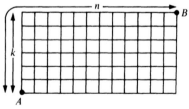

Example *Let k and n be integers with $1 \leq k \leq n$. Show that*

$$\binom{n}{k} = \binom{n-1}{k} + \binom{n-1}{k-1}.$$

Solution There are many ways of proving this identity. The most boring would be to express each side in terms of factorials and to show that you get the same answers. Another way would be to see how many x^k you get in $(1 + x)^n$ and in $(1 + x)(1 + x)^{n-1}$ (which are the same thing). But the solution we give uses the routes in a grid as discussed above.

There are $\binom{n}{k}$ shortest routes from A to B in the grid illustrated. These routes fall into two separate types, those whose final move is 'right' (i.e. those which approach B from B_1) and those whose final move is 'up' (i.e. those which approach B from B_2). Hence

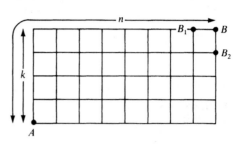

$$\binom{n}{k} = \text{number of routes from } A \text{ to } B$$

$$= (\text{number of routes from } A \text{ to } B_1) + (\text{number of routes from } A \text{ to } B_2)$$

$$= \binom{n-1}{k} + \binom{n-1}{k-1}$$

as required. □

The identity in that example is the basis of *Pascal's triangle* of the binomial coefficients, where each is obtained by adding two earlier ones:

$$
\begin{array}{c}
1 \\
1 \quad 1 \\
1 \quad 2 \quad 1 \\
1 \quad 3 \quad 3 \quad 1 \\
1 \quad 4 \quad 6 \quad 4 \quad 1 \\
1 \quad 5 \quad 10 \quad 10 \quad 5 \quad 1 \\
\vdots
\end{array}
$$

Example *How many solutions are there of the equation*

$$x_1 + x_2 + x_3 + x_4 = 6$$

where each x_i is a non-negative integer? How many solutions are there of the equation

$$x_1 + x_2 + \cdots + x_k = n$$

where each x_i is a non-negative integer?

The binomial coefficients

Solution Again there are many ways of tackling this question (one of which we'll see in the exercises). But as before we choose a solution using routes in a grid. For reasons which will soon become apparent we choose a grid with four horizontal roads and six moves 'right'.

Two shortest routes from A to B are illustrated. If you then regard the number of moves in row i as x_i then these routes translate to the given solutions of the equation $x_1 + x_2 + x_3 + x_4 = 6$.

There is clearly a one-to-one correspondence between the routes and the solutions and it follows that the number of solutions is $\binom{9}{3} = 84$. Similar reasoning shows that the number of solutions of

$$x_1 + x_2 + \cdots + x_k = n$$

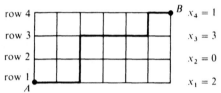

equals the number of shortest routes from corner to corner in a grid with k horizontal roads and n moves 'right'. In such a route in this grid there are $n + k - 1$ moves of which any $k - 1$ must be 'up'. Hence the number of solutions is $\binom{n+k-1}{k-1}$. □

We have now considered the *bi*nomial coefficients where, as the prefix 'bi' implies, we are involved with a two-way choice, namely to include or not to include a particular item. We can easily extend this idea to *multi*nomial coefficients.

Example *How many different ten-figure numbers can be formed by writing down the ten digits* 4, 4, 4, 4, 3, 3, 3, 2, 2 *and* 1 *in some order?*

Solution The answer is

$$\frac{10!}{4! \times 3! \times 2!} = 12\,600.$$

To see this we can imitate the solution to our very first example on binomial coefficients. Imagine the long list of 10! numbers formed by the given 10 digits in any order: this list will have many repeats. For example the number

$$4\,231\,442\,343$$

will be in the list $4! \times 3! \times 2!$ times because shuffling around the 4s within their four places, the 3s within their places and the 2s within theirs will not change the number. Similarly each possible number will be in the list $4! \times 3! \times 2!$ times, and the result follows.

Alternatively note that the problem is equivalent to choosing four positions from the ten available for the 4s, then three positions from the remaining six for the 3s, and two positions from the remaining three for the 2s: the position of the 1 is then determined. The number of ways of doing this is

$$\binom{10}{4}\binom{6}{3}\binom{3}{2}\binom{1}{1} = 12\,600$$

as before. □

That answer,

$$\frac{10!}{4!\,3!\,2!\,1!} = \binom{10}{4}\binom{6}{3}\binom{3}{2}\binom{1}{1},$$

can also be thought of as the number of ways of placing ten items into four boxes with four in the first box, three in the second, two in the third, and one in the fourth. In general, if $r \geqslant 2$ and $k_1, k_2, \ldots, k_r$ are integers with $k_1 + k_2 + \cdots + k_r = n$, then the number of ways of placing n items in r boxes with k_1 items in the first box, k_2 in the second, ... and k_r in the rth is called a *multinomial coefficient* and is denoted by

$$\binom{n}{k_1 \quad k_2 \quad \cdots \quad k_r}.$$

If any of the integers k_i is negative then the coefficient is zero, but if all the k_i are non-negative then either of the arguments used in the last solution leads to

$$\binom{n}{k_1 \quad k_2 \quad \cdots \quad k_r} = \frac{n!}{k_1!\,k_2!\cdots k_r!}.$$

We only use multinomial coefficients in the cases when $r \geqslant 2$ and when the r numbers in the bottom row add to the number in the top row. In particular, notice that when $r = 2$ the multinomial coefficients reduce to the usual binomial coefficient since

$$\binom{n}{k_1 \quad k_2} = \frac{n!}{k_1!\,k_2!} = \frac{n!}{k_1!\,(n-k_1)!} = \binom{n}{k_1}.$$

Also some of the properties of the binomial coefficients easily extend to the multinomial coefficients.

Example *Show that*

$$\binom{n}{k_1 \quad k_2 \quad k_3 \quad \cdots \quad k_r} = \binom{n-1}{k_1-1 \quad k_2 \quad k_3 \quad \cdots \quad k_r}$$

$$+ \binom{n-1}{k_1 \quad k_2-1 \quad k_3 \quad \cdots \quad k_r}$$

$$+ \cdots + \binom{n-1}{k_1 \quad k_2 \quad k_3 \quad \cdots \quad k_r-1}.$$

The binomial coefficients

Solution We have to place n items in r boxes with k_1 items in the first box, k_2 in the second, ... and k_r in the rth. If the first item is placed in box i then the remaining $n - 1$ items must be placed in the r boxes with k_1 items in the first box, k_2 in the second, ..., $k_i - 1$ in the ith, ... and k_r in the rth. Since i can take any value from 1 to r the stated result follows easily. □

Example *The binomial expansion can be written in the form*

$$(a+b)^n = \binom{n}{n\ 0}a^n + \binom{n}{n-1\ 1}a^{n-1}b + \binom{n}{n-2\ 2}a^{n-2}b^2 + \cdots + \binom{n}{0\ n}b^n$$

$$= \sum_{\substack{k_1,k_2 \geq 0 \\ k_1+k_2=n}} \binom{n}{k_1\ k_2} a^{k_1} b^{k_2}.$$

Show that in general

$$(a_1 + a_2 + \cdots + a_r)^n = \sum_{\substack{k_1,\ldots,k_r \geq 0 \\ k_1+\cdots+k_r=n}} \binom{n}{k_1\ \cdots\ k_r} a_1^{k_1} a_2^{k_2} \cdots a_r^{k_r}.$$

Solution Again we can imitate the method used for the binomial expansion:

$$(a_1 + a_2 + \cdots + a_r)^n = \underbrace{(a_1 + a_2 + \cdots + a_r)(a_1 + a_2 + \cdots + a_r) \cdots (a_1 + a_2 + \cdots + a_r)}_{n \text{ lots}}.$$

When multiplying out these brackets the coefficient of $a_1^{k_1} a_2^{k_2} \cdots a_r^{k_r}$ will equal the number of ways of choosing a_1 from k_1 of the brackets, choosing a_2 from k_2 of the remaining brackets, etc., and the result follows. □

We conclude this introductory chapter with a large selection of exercises: for those marked [H] a helpful (?) hint will be found in the section which starts on page 224 and for those marked [A], which require a numerical answer, the actual answer is given in the section which begins on page 249.

Exercises

1. Show by various methods that

$$\binom{n}{k} = \binom{n}{n-k}.$$ [H]

2. There are n people in a queue for the cinema (and, being in England, the order of people in the queue never changes). They are let into the cinema in k batches, each batch consisting of one or more persons. In how many ways can the k batches be chosen? [H,A]

3. How many solutions are there of the equation
$$x_1 + x_2 + \cdots + x_k = n$$
where each x_i is a positive integer? [H,A]

4. (i) Use the answer to exercise 3 to find an alternative verification of the fact that the number of solutions of
$$x_1 + x_2 + \cdots + x_k = n$$
with each x_i a non-negative integer is $\binom{n+k-1}{k-1}$. [H]

(ii) By considering the number of x_i which are zero in (i) and by using the answer to exercise 3 for the positive x_i, show that
$$\binom{n+k-1}{k-1} = \binom{k}{0}\binom{n-1}{k-1} + \binom{k}{1}\binom{n-1}{k-2} + \binom{k}{2}\binom{n-1}{k-3} + \cdots + \binom{k}{k-1}\binom{n-1}{0}.$$

5. In a row of n seats in the doctor's waiting-room k patients sit down in a particular order from left to right. They sit so that no two of them are in adjacent seats. In how many different ways could a suitable set of k seats be chosen? [H,A]

6. By considering colouring k out of n items using one of two given colours for each item, show that
$$\binom{n}{0}\binom{n}{k} + \binom{n}{1}\binom{n-1}{k-1} + \binom{n}{2}\binom{n-2}{k-2} + \cdots + \binom{n}{k}\binom{n-k}{0} = 2^k \binom{n}{k}.$$

7. Show by each of the following methods that
$$\binom{n}{0}^2 + \binom{n}{1}^2 + \binom{n}{2}^2 + \cdots + \binom{n}{n}^2 = \binom{2n}{n}.$$

(i) Use the expansion of $(1 + x)^{2n}$. [H]
(ii) Consider the choice of n people from a set of $2n$ which consists of n men and n women. [H]
(iii) Count the number of routes in a suitable grid. [H]

8. (i) There is a group of $2n$ people consisting of n men and n women from whom I wish to choose a subset, the only restriction being that the number of men chosen equals the number of women chosen. Use the result of exercise 7 to show that the subset may be chosen in $\binom{2n}{n}$ ways.

(ii) I now wish to choose the subset as in (i) and then to choose a leader from the men in the subset and a leader from the women in the subset. Calculate the number of ways of choosing the subset first and then the leaders, and also calculate the number of ways of choosing the leaders first and then the remainder of the subset. Deduce that

$$1^2\binom{n}{1}^2 + 2^2\binom{n}{2}^2 + 3^2\binom{n}{3}^2 + \cdots + n^2\binom{n}{n}^2 = n^2\binom{2n-2}{n-1}.$$

(iii) Show by various methods that

$$\binom{n}{1} + 2\binom{n}{2} + 3\binom{n}{3} + \cdots + n\binom{n}{n} = n2^{n-1}. \qquad [H]$$

9. (i) If you draw n straight lines in the plane, no two of them being parallel and no three of them meeting at a point, how many intersection points will there be altogether? [A]

(ii) If you draw n straight lines in the plane consisting of x_1 parallel in one direction, x_2 parallel in a different direction, ... and x_k parallel in another different direction, and with no three of the lines meeting at a point, show that the number of intersection points is

$$\tfrac{1}{2}(n^2 - (x_1^2 + x_2^2 + \cdots + x_k^2)). \qquad [H]$$

(iii) Draw a collection of 17 straight lines, no three meeting at a point, with a total of 101 intersection points. [H,A]

10. (i) How many different rectangles can be seen in an $n \times n$ grid like the one shown? (One such rectangle is highlighted: of course the rectangles must be at least one box wide and deep, and squares are allowed.) [H,A]

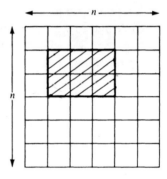

(ii) Use the answer to (i) to show that the number of rectangles which lie in the top left-hand $r \times r$ corner of the grid and which touch one of the two internal boundary lines of that corner (like the lower rectangle shown) is r^3. [H]

(iii) Show that there are

$$1^3 + 2^3 + 3^3 + \cdots + n^3$$

rectangles in the grid and deduce that

$$1^3 + 2^3 + 3^3 + \cdots + n^3 = (\tfrac{1}{2}n(n+1))^2.$$

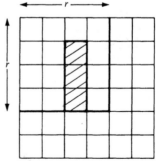

(iv) How many of the $1^3 + 2^3 + 3^3 + \cdots + n^3$ rectangles are squares? [A]

11. (i) Show that the sum of the numbers in the rth row of the array illustrated at top of next page is $\binom{r+1}{2}$ and hence express the grand total of numbers in the array as a single binomial coefficient. [A]

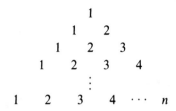

(ii) By adding up the numbers in the array in a different way deduce that

$$1 \cdot n + 2 \cdot (n-1) + 3 \cdot (n-2) + \cdots + n \cdot 1 = \binom{n+2}{3}.$$

(iii) How many triangles oriented the same way up as *ABC* can be seen in a grid like the one shown but consisting of *n* rows? [H,A]

(iv) Express the number of triangles which can be seen the other way up as a sum of binomial coefficients. [H,A]

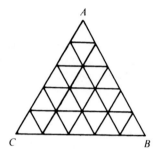

12. If *n* points are placed around the circumference of a circle and each pair of points is joined by a straight line, then the circle is divided into a number of regions. Show that, if no three of the lines meet inside the circle, then the number of regions will be

$$1 + \binom{n}{2} + \binom{n}{4} \qquad [H]$$

13. At an ice-cream stall *m* + *n* people are in a queue to buy one 50p ice-cream each. The stall-keeper has no change. Precisely *m* of the people have just a 50p coin and the remaining *n* people have just a £1 coin. By considering routes in a grid, find the probability that the stall-keeper always has enough change to serve the customers in turn. [H,A]

2
How many trees?

Graph theory will be another central theme occurring throughout this book, but in this introductory chapter we are only going to study in detail one special type of graph and to use some of our techniques from the first chapter to count them.

A *graph* $G = (V, E)$ consists of a finite non-empty set of *vertices* V and a set of *edges* E which is any subset of the pairs $\{\{v, w\}: v, w \in V, v \neq w\}$. A particular edge $\{v, w\}$ is usually written vw (or wv) and is said to *join* the vertex v to the vertex w, and v, w are called the *endpoints* of the edge vw.

Example *Let $G = (V, E)$ be the graph with set of vertices (or 'vertex-set') $V = \{Meryl\ Streep,\ Dustin\ Hoffman,\ Jeremy\ Irons,\ Anne\ Bancroft,\ Robert\ Redford\}$ and let the 'edge-set' be defined by*

$$E = \{vw: v, w \in V \text{ made a film together before 1992}\}.$$

So, for example, Meryl Streep is joined to Jeremy Irons because of The French Lieutenant's Woman. We can illustrate G by representing the vertices as points in the plane and by joining two vertices by a line (or curve) if they are joined by an edge in the graph:

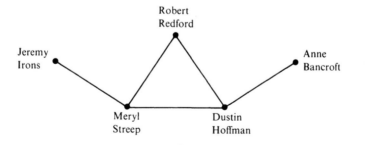

Note that in a graph any pair of distinct vertices may or may not be joined by an edge: we do not allow a vertex to be joined to itself nor do we allow two vertices to be joined by several edges. Some authors refer to our graphs as 'simple graphs': these simple structures are sufficient for many applications.

Before we can proceed to study the many facets of graph theory we need some technical terms. In a graph a *path* (or *trail*) is a sequence of vertices

$$v_1, v_2, v_3, \ldots, v_{n-1}, v_n$$

(where repeats are allowed) such that $v_1v_2, v_2v_3, \ldots, v_{n-1}v_n$ are all different edges of the graph. A path of the form

$$v_1, v_2, \ldots, v_{n-1}, v_n, v_1$$

where $n > 1$ and the first n vertices are also all different is called a *cycle* (or *circuit*). So, in our pictures of graphs, 'paths' take you from one vertex to another along edges: you may cross over a route you took earlier but you must not use any stretch of road more than once. A 'cycle' is a round walk never visiting any point more than once *en route*.

Example (*based on part of the Birmingham canal network*)

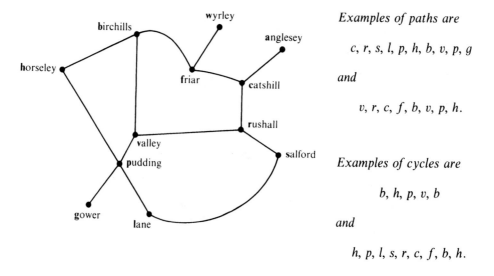

Examples of paths are

$$c, r, s, l, p, h, b, v, p, g$$

and

$$v, r, c, f, b, v, p, h.$$

Examples of cycles are

$$b, h, p, v, b$$

and

$$h, p, l, s, r, c, f, b, h.$$

□

Naturally enough the graph in that example has the property that it is possible to get from any vertex to any other along a path: such a graph is called *connected*. If this fails to happen the graph is called *disconnected*: in that case the graph falls into several 'pieces' or 'components'. (To define this formally we need the idea of a *subgraph* of $G = (V, E)$, which is simply another graph whose vertex-set is a subset of V and whose edge-set is a subset of E. Then a *component* of a graph is a maximal connected subgraph of it.)

In this chapter we are going to study 'trees', which are a special type of graph. You have probably already met the word 'tree' used to describe some graph-like structures:

Examples

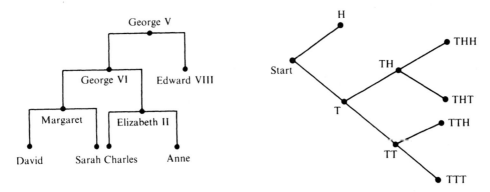

Part of the royal family tree

*A probability tree:
tossing a coin up to three times, stopping
sooner if number of heads exceeds tails*

Notice that those structures have two common properties: firstly they are in one piece and secondly if you trace a route through the structure you always come to a dead-end. It is precisely these two properties which characterise 'trees': a *tree* is a connected graph without any cycles.

Example

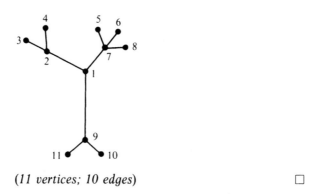

(11 vertices; 10 edges)

We commented above that any route in a tree will eventually come to a dead-end. In general, in a graph $G = (V, E)$ the *degree* of a vertex $v \in V$ is the number of vertices to which v is joined, and it is written δv. So, for instance, the vertex 7 in the above example has degree $\delta 7 = 4$, and the sum of all the degrees is 20, which is twice the number of edges. In this new terminology the 'dead-ends' are simply those vertices of degree 1. If a tree has more than one vertex then imagine a 'longest' path possible; i.e. one using the most edges. It is not difficult to see that if this path is from vertex v to vertex w then v and w both have degree 1. In the exercises we shall use this fact to show that a tree with n vertices has precisely $n-1$ edges and that the sum of the degrees of its n vertices is therefore $2n-2$.

We are now ready to count the number of trees on vertex-set $\{1, 2, \ldots, n\}$ with the degree of each vertex prescribed.

Example *How many trees are there on vertex-set $\{1, 2, 3, 4, 5\}$ with the degrees of the vertices given by $\delta 1 = 3$, $\delta 2 = 2$, $\delta 3 = 1$, $\delta 4 = 1$ and $\delta 5 = 1$?*

Solution There are only three such trees. In order to find them systematically note that the vertex 5 must be joined to just one other. Imagine that it is joined to vertex i and picture the trees obtained by deleting the vertex 5 and the edge $5i$:

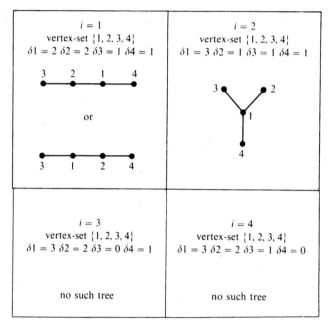

From these smaller trees it is then easy to construct the required trees on vertex-set $\{1, 2, 3, 4, 5\}$ by adding the extra edge $5i$:

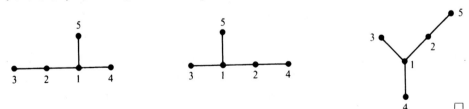

We shall now use that general principle to count trees with any prescribed degrees:

Theorem *Let $n \geq 2$ and let $d_1, \ldots, d_n$ be positive integers with sum $2n - 2$. Then the number of trees on vertex-set $\{1, \ldots, n\}$ with the degrees of the vertices given by*

$\delta 1 = d_1, \ldots, \delta n = d_n$ is

$$\frac{(n-2)!}{(d_1 - 1)!\, (d_2 - 1)! \cdots (d_n - 1)!}.$$

Proof Note first that $(d_1 - 1) + \cdots + (d_n - 1) = n - 2$ and that the given answer is the multinomial coefficient

$$\binom{n-2}{d_1 - 1 \quad d_2 - 1 \quad \cdots \quad d_n - 1}.$$

In fact the theorem is equivalent to the result that if $d_1, \ldots, d_n$ are *any* integers (not necessarily positive) which add to $2n - 2$ then the number of trees on vertex-set $\{1, \ldots, n\}$ with degrees $d_1, \ldots, d_n$ is that multinomial coefficient. For if any d_i is zero or less then the number of trees and the given multinomial coefficient are both zero.

We now prove the theorem by induction on n, the case $n = 2$ being trivial. For in that case $d_1 = d_2 = 1$, there is just one such tree, and the given multinomial coefficient equals 1 in this case.

So assume that $n > 2$, that the $d_1, \ldots, d_n$ are positive integers which add to $2n - 2$ and that the result is known for smaller n. Since the n positive integers $d_1, \ldots, d_n$ add up to less than $2n$, it follows that one of them (d_n say) is 1. So in the trees which we are trying to count the vertex n will be joined to exactly one other vertex, i say. Imagine (as we did in the previous example) the trees obtained by deleting the vertex n and the edge ni. Such a tree will have vertex-set $\{1, \ldots, n-1\}$ and degrees $\delta 1 = d_1, \ldots, \delta i = d_i - 1, \ldots$ and $\delta(n-1) = d_{n-1}$. Hence, letting i take all possible values, we obtain

$$\boxed{\begin{array}{c}\text{number of trees on vertex-}\\\text{set }\{1, \ldots, n\}\text{ with}\\ \delta 1 = d_1, \delta 2 = d_2, \ldots, \delta n = d_n\end{array}} = \sum_{i=1}^{n-1} \boxed{\begin{array}{c}\text{number of trees on vertex-}\\\text{set }\{1, \ldots, n-1\}\text{ with}\\ \delta 1 = d_1, \ldots, \delta i = d_i - 1, \ldots, \delta(n-1) = d_{n-1}\end{array}}.$$

How many trees are there on vertex-set $\{1, \ldots, n-1\}$ with $\delta 1 = d_1, \ldots, \delta i = d_i - 1$, $\ldots$ and $\delta(n-1) = d_{n-1}$? Since $d_1, \ldots, d_i - 1, \ldots, d_{n-1}$ add to $2(n-1) - 2$ the induction hypothesis and our opening comments tell us that the number of such trees is

$$\binom{(n-1)-2}{d_1 - 1 \quad \cdots \quad d_i - 2 \quad \cdots \quad d_{n-1} - 1}.$$

Hence

$$\boxed{\begin{array}{c}\text{number of trees on vertex-}\\\text{set }\{1, \ldots, n\}\text{ with}\\ \delta 1 = d_1, \delta 2 = d_2, \ldots, \delta n = d_n\end{array}} = \sum_{i=1}^{n-1} \binom{n-3}{d_1 - 1 \quad \cdots \quad d_i - 2 \quad \cdots \quad d_{n-1} - 1}.$$

But the summation property of the multinomial coefficients established in the example on page 8 shows that this sum is given by

$$\binom{n-2}{d_1-1 \quad \cdots \quad d_{n-1}-1} = \frac{(n-2)!}{(d_1-1)!\cdots(d_{n-1}-1)!}$$

$$= \frac{(n-2)!}{(d_1-1)!\cdots(d_{n-1}-1)!\underbrace{(d_n-1)!}_{=1}}$$

and the result follows for n. Hence the proof by induction is complete. □

We now use that theorem to count the total number of trees on vertex-set $\{1, \ldots, n\}$: the delightful result is named after Arthur Cayley who first stated it (and proved it in the case $n = 6$) in 1889. More details of the history of the theorem can be found in the fascinating book *Graph theory 1736–1936* quoted in the bibliography.

Theorem (Cayley) *For $n \geq 2$ the number of trees on vertex-set $\{1, \ldots, n\}$ is n^{n-2}.*

Proof Note firstly that the multinomial expansion, derived in the last example of the previous chapter, tells us that

$$(a_1 + \cdots + a_n)^{n-2} = \sum_{\substack{k_1, \ldots, k_n \geq 0 \\ k_1 + \cdots + k_n = n-2}} \binom{n-2}{k_1 \quad \cdots \quad k_n} a_1^{k_1} \cdots a_n^{k_n}$$

$$= \sum_{\substack{d_1, \ldots, d_n \geq 1 \\ (d_1-1)+\cdots+(d_n-1)=n-2}} \binom{n-2}{d_1-1 \quad \cdots \quad d_n-1} a_1^{d_1-1} \cdots a_n^{d_n-1}.$$

In particular, putting each a_i equal to 1 gives

$$n^{n-2} = (1 + 1 + \cdots + 1)^{n-2} = \sum_{\substack{d_1, \ldots, d_n \geq 1 \\ d_1+\cdots+d_n=2n-2}} \binom{n-2}{d_1-1 \quad \cdots \quad d_n-1}.$$

Now to calculate the number of trees on vertex-set $\{1, \ldots, n\}$ we must add up the numbers of trees whose vertices have degrees $d_1, \ldots, d_n$ over all possible values of the d_i. The $d_1, \ldots, d_n$ must be positive integers which add up to $2n - 2$. So from the previous theorem we have

$$\boxed{\begin{array}{c}\text{number of trees} \\ \text{on vertex-set} \\ \{1, \ldots, n\}\end{array}} = \sum_{\substack{d_1, \ldots, d_n \geq 1 \\ d_1+\cdots+d_n=2n-2}} \boxed{\begin{array}{c}\text{number of trees on vertex} \\ \text{set } \{1, \ldots, n\} \text{ with} \\ \delta 1 = d_1, \delta 2 = d_2, \ldots, \delta n = d_n\end{array}}$$

$$= \sum_{\substack{d_1, \ldots, d_n \geq 1 \\ d_1+\cdots+d_n=2n-2}} \binom{n-2}{d_1-1 \quad \cdots \quad d_n-1}$$

$$= n^{n-2}$$

and Cayley's theorem is proved. □

How many trees?

Example *There are 16 trees on vertex-set $\{1, 2, 3, 4\}$. There are also 16 pairs of integers of the form ab where a and b are freely chosen from $\{1, 2, 3, 4\}$: we can therefore 'name' these 16 trees using the 16 pairs. The reason for the choice of names will soon be made clear.*

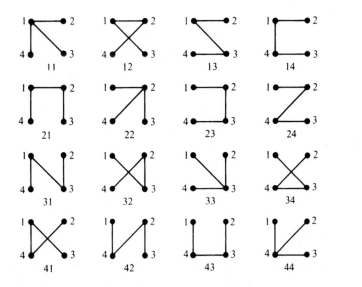

The very neat answer of n^{n-2} for the number of trees on n particular vertices has led to a wide variety of interesting proofs of Cayley's theorem. One proof, due to Heinz Prüfer in 1918, showed a one-to-one correspondence between the collection of trees on vertex-set $\{1, \ldots, n\}$ and the collection of ordered lists of $n - 2$ numbers chosen freely (i.e. with repeats allowed) from that same set. His method enables us to use these lists to 'name' each tree, as we did in the previous example. We now give Prüfer's algorithm for naming a tree: the tree's name is sometimes referred to as its 'Prüfer code'.

Algorithm *Given a tree T on vertex-set $\{1, \ldots, n\}$, to find its Prüfer code:*

   I. *Find the lowest vertex of degree 1, v say. Let w be the vertex joined to v.*
  II. *Write down w and delete the vertex v and the edge vw.*
 III. *If you are left with a tree of more than one edge go to* I: *otherwise stop.*

Then the resulting list of numbers written down is the Prüfer code *of T.*

Since the tree T has n vertices and the algorithm writes down one number each time a vertex is deleted until just two vertices remain, it follows that the resulting list will certainly be of $n - 2$ numbers from $\{1, \ldots, n\}$.

Example *Applying the algorithm to the tree T illustrated gives its Prüfer code as 2332:*

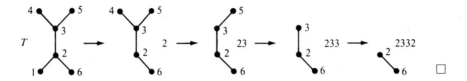

Note that in the given example each vertex v occurs in the code $\delta v - 1$ times. It is clear that this will always be the case since the algorithm writes down v each time an edge leading from it is deleted until v remains with degree 1.

Theorem *Every list of $n - 2$ numbers freely chosen from the set $\{1, \ldots, n\}$ is the Prüfer code of precisely one tree on vertex-set $\{1, \ldots, n\}$.*

Proof There are n^{n-2} trees on vertex-set $\{1, \ldots, n\}$ and there are n^{n-2} lists of $n - 2$ numbers chosen freely from that set. Therefore, if we show that the given algorithm gives rise to different codes for different trees, then it follows that each such list is the Prüfer code of precisely one tree.

So assume that trees T and T' on vertex-set $\{1, \ldots, n\}$ both have Prüfer code $a_1 a_2 \cdots a_{n-2}$: we shall aim to show that $T = T'$. By our comments above, in general each vertex v occurs in the code $\delta v - 1$ times. Since T and T' have the same codes it follows that the degree of each vertex v in T is the same as its degree in T'. Now, since the algorithm wrote down a_1 first in each case, it follows that T and T' are of the following form:

We shall show that $i = j$. If not then $i < j$, say, and as j is the lowest vertex of degree 1 in T' it follows that the vertex i (being lower than j) has degree greater than 1 in T'. But the vertex i has degree 1 in T which contradicts our above comments about the degrees: hence $i = j$ as claimed.

The algorithm then continues with the edge $a_1 i$ removed in each case and we could apply a similar argument to each of these reduced trees. Continuing in this way it is clear that T and T' have identical edges and are the same tree, as required. □

If I gave you the Prüfer code of a tree would you be able to reconstruct the tree from it? In that last proof we began to see how such a reconstruction would work: if the code is $a_1 \cdots a_{n-2}$ then the tree has to have an edge $a_1 i$ where i is the lowest vertex of degree 1. Since each vertex v occurs in the code $\delta v - 1$ times it follows that i is the lowest vertex not occurring in the code. That key observation enables us to define an algorithm which performs the inverse process to the previous one; i.e. it starts with the Prüfer code and constructs the tree.

Algorithm *Given a list $a_1 \cdots a_{n-2}$ of numbers freely chosen from $\{1, \ldots, n\}$, to find the tree T with that list as its Prüfer code:*

  I. *Write down a first list $a_1 \cdots a_{n-2}$, a second list $1, 2, \ldots, n$ and start with vertex-set $\{1, \ldots, n\}$ and an empty set of edges.*
 II. *Find the lowest number, i say, which is in the second list but is not in the first. Delete the leading entry in the first list, j say, delete i from the second list and add the edge ij to the edge-set.*
III. *If there is still any number left in the first list go to II. Otherwise, if the first list is empty, then the second list will consist of just two numbers. Add one last edge to the edge-set consisting of those two numbers joined together: then stop.*

Example *To reconstruct the tree T with Prüfer code 2332 (as the earlier example):*

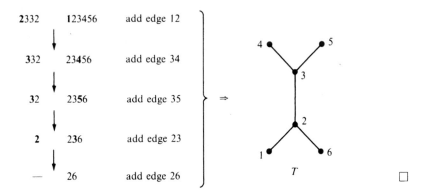

When counting the trees on n vertices we have labelled the vertices $1, 2, \ldots, n$. The problem of counting the trees if the vertices are unlabelled is a completely different (and, for large n, somewhat harder) one. For example in the list of 16 trees on vertex-set $\{1, 2, 3, 4\}$ there are essentially only two different structures, namely

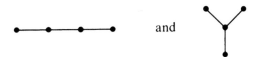

In most of our future work on graphs the actual labels assigned to the vertices will be irrelevant.

Example Let $G' = (V', E')$ be the graph consisting of vertices

$$V' = \{France, Germany, Spain, Switzerland, Netherlands\}$$

and let the set of edges be defined by

$$E' = \{vw : v, w \text{ have a common border}\}.$$

Then G' can be illustrated as follows:

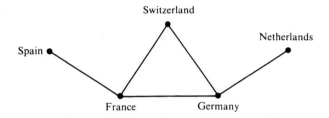

If we ignore the labels on the vertices this graph has the same structure as the film-buff example $G = (V, E)$ which opened the chapter:

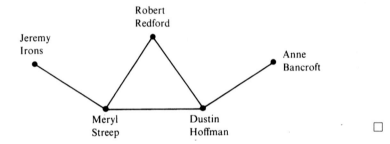

Formally these two graphs are *isomorphic* (i.e. there is a bijection from V to V' – taking v = Robert Redford to v' = Switzerland for example – so that vw is an edge of G if and only if $v'w'$ is an edge of G'). Another example of isomorphic graphs is found in the family tree and the probability tree in the earlier example on page 15.

These ideas need not concern us unduly: in future chapters we shall generally only be concerned with the structure of the graph, not with the specific labels of the vertices.

Examples The complete graph K_n consists of n vertices each pair of which is joined by an edge:

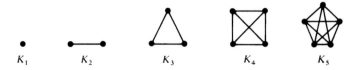

The complete bipartite graph $K_{m,n}$ consists of $m + n$ vertices and mn edges so that each of m of the vertices is joined to each of the remaining n vertices:

$K_{1,3}$ $K_{2,3}$ $K_{2,4}$ $K_{3,3}$ □

Exercises

1. (i) How many edges are there in the complete graph K_n? [A]
   (ii) How many graphs are there with vertex-set $\{1, \ldots, n\}$ and with precisely m edges? [H,A]
   (iii) How many graphs are there altogether on vertex-set $\{1, \ldots, n\}$? [H,A]

2. (The handshaking lemma) Show that in any graph $G = (V, E)$
$$\sum_{v \in V} \delta v = 2|E|,$$
where $|E|$ denotes the number of edges. (In general we use $|X|$ to denote the number of elements in the set X.) [H]

3. Let $G = (V, E)$ be a graph in which $E \neq \emptyset$ and in which each vertex has even degree. By considering a non-trivial component (in which each vertex therefore has degree at least 2) show that G has a cycle. Show also that, if the edges used in this cycle are removed from the graph, then in the new graph each vertex still has even degree. Deduce that G has a collection of cycles which between them use each edge in E exactly once.

4. (i) Prove that, for any tree $G = (V, E)$, $|E| = |V| - 1$. [H]
   (ii) Show also that if $G = (V, E)$ is a connected graph with $|E| = |V| - 1$ then G is a tree. [H]

5. For $n \geq 2$ how many of the n^{n-2} trees on vertex-set $\{1, 2, \ldots, n\}$ have
   (i) a vertex of degree $n - 1$? [A]
   (ii) a vertex of degree $n - 2$? [H,A]
   (iii) all the vertices of degree 1 or 2? [H,A]
   (iv) the degree of vertex 1 equal to 1? [H,A]

   Show that the proportion of these n^{n-2} trees with $\delta 1 = 1$ tends to $1/e$ as n tends to infinity. [H]

6. Hydrocarbon molecules consist of carbon atoms (of valency 4) and hydrogen atoms (of valency 1) and they can be illustrated as connected graphs. For example, molecules of butane and 2-methylpropane (isobutane) both contain four carbon atoms and ten hydrogen atoms as shown:

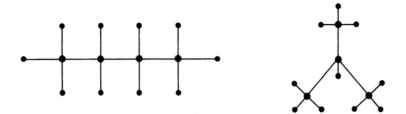

These two non-isomorphic structures with the same chemical formula C_4H_{10} are examples of *isomers*.

(i) Show that any hydrocarbon molecule with formula of the form C_nH_{2n+2} (called a *paraffin* or *alkane*) has a structure which is a tree, but that those molecules with a formula of the form C_nH_{2n} (the *alkenes*) do not have a tree structure. [H]

(ii) Show that the two examples above are the only isomers of C_4H_{10} and find the maximum possible number of isomers of C_3H_8 and of C_5H_{12}. [H,A]

7. Show that, for a graph $G = (V, E)$ with k components,
$$|V| - k \leq |E| \leq \tfrac{1}{2}(|V| - k)(|V| - k + 1).$$
[H]

3

The marriage theorem

Whereas our previous chapters have been concerned with *counting* various selections, here we ask about the *existence* of certain types of selections of elements from sets. The work centres around a famous theorem due to Philip Hall in 1935 and it is easiest to understand when presented in its 'marriage' version, where the selection takes the form of choosing wives for husbands (or vice versa). But we shall then deduce various other forms of the theorem and, in later chapters, see several applications of it.

Example In a group of seven boys and six girls of marriageable age

> girl 1 knows boys 1', 2' and 3',
> girl 2 knows boys 2' and 3',
> girl 3 knows boys 3', 5' and 7',
> girl 4 knows boys 1' and 2',
> girl 5 knows boys 1', 2' and 3',
> girl 6 knows boys 4', 5' and 6'.

Is it possible to find each of the girls a husband (i.e. a different boy for each from amongst those whom she knows)?

Solution It is impossible to find each of the girls a husband. One way of seeing this is to note that the four girls 1, 2, 4 and 5 only know three boys between them, namely 1', 2' and 3': so there's clearly no hope of finding all four of those girls a husband. □

Clearly, to have any hope of finding each girl a husband, it must be the case that any subset of the girls (r of them, say) know between them at least r boys. In fact it turns out that husbands can be found **if and only if** this condition holds.

Example
    Girl 1 knows boys 1' and 3',
    girl 2 knows boys 2' and 3',
    girl 3 knows boys 1', 3', 4' and 5',
    girl 4 knows boys 2', 4', 6' and 7,
    girl 5 knows boys 1' and 5',
    girl 6 knows boys 1' and 2'.

In this case any set of girls know between them at least as many boys. For example the girls $\{1, 2, 6\}$ know between them the boys $\{1', 2', 3'\}$. Can we find a husband for each girl from the boys whom they know?

Solution It's easy to find husbands for the girls by trial and error, but we'll apply a process which will in fact work in general (as we shall see in the proof of the next theorem). We start to choose different boys for each of the girls in any way we like until we reach a girl for whom there is no boy left to choose. For example 1 could get engaged to 1', 2 to 2', 3 to 3', 4 to 4' and 5 to 5'. But then girl 6 only knows 1' and 2', and they are already engaged. How can she find a partner?

    Girl 6 throws a party. She invites all the boys she knows: they invite their fiancées: those girls invite all the boys they know who haven't already been invited: those boys invite their fiancées: those girls invite all the boys they know who haven't already been invited: . . . . This process continues until some boy (B' say) is invited who is not already engaged. In this case it leads to

$$\{6\} \text{ invites } \{1', 2'\} \text{ invites } \{1, 2\} \text{ invites } \{3'\} \text{ invites}$$
$$\{3\} \text{ invites } \{4', 5'\} \text{ invites } \{4, 5\} \text{ invites } \{6', 7'\}$$

We stop there because, for example, $B' = 7'$ is not engaged. Now B' (7') dances with a girl he knows who invited him (4). Her fiancé (4') is a bit annoyed so he dances with a girl who invited him (3): her fiancé (3') is a bit put out so he dances with a girl who invited him (1): her fiancé (1') is a bit miffed so he dances with a girl who invited him (6). In effect we are working back through the sets listed above:

$$\{6\}\text{–}\{1', 2'\}\text{–}\{1, 2\}\text{–}\{3'\}\text{–}\{3\}\text{–}\{4', 5'\}\text{–}\{4, 5\}\text{–}\{6', 7'\}$$

    Some engaged couples (1, 1'; 3, 3'; 4, 4') together with girl 6 and boy 7' paired off into new couples

    This smoochy dance is so successful that the dancing couples break off any former engagements and get engaged. No other couples are affected. That leaves

    1 engaged to 3': 2 to 2': 3 to 4': 4 to 7': 5 to 5' and 6 to 1'

and we have managed to find husbands for all the girls.   □

The marriage theorem

Theorem (Hall's theorem – marriage form) *A set of girls can all choose a husband each from the boys that they know if and only if any subset of the girls (r of them, say) know between them at least r boys.*

Proof ($\Rightarrow$) If the girls can all find husbands, then clearly any r girls must know between them at least r boys, namely their husbands.

($\Leftarrow$) Let the girls be named $1, 2, \ldots, n$ and assume that any subset of them (r say) know between them at least r boys. We shall show by induction on m ($1 \leq m \leq n$) that the girls $1, 2, \ldots, m$ can all find husbands from the boys that they know.

The case $m = 1$ is trivial since the girl 1 knows at least one boy whom she can then choose as her husband. So assume that $m \geq 1$, that the girls $1, 2, \ldots, m$ have found fiancés (named $1', 2', \ldots, m'$ respectively) and that we now wish to find fiancés for the girls $1, 2, \ldots, m$ and $m + 1$. These $m + 1$ girls know between them at least $m + 1$ boys and only m of those boys are engaged. So it is tempting to say that the girl named $m + 1$ must know some boy who is not engaged: if she does then we can find her a fiancé and the proof is complete. But (and this is what makes the result non-trivial) even though the girls $1, 2, \ldots, m, m + 1$ know between them at least $m + 1$ boys, the girl $m + 1$ may personally only know boys from amongst $1', 2', \ldots, m'$, all of whom are already engaged. So how does she find herself a fiancé?

As in the previous example, let girl $m + 1$ throw a party. She invites all the boys she knows: they invite their fiancées: those girls invite all the boys they know who haven't already been invited: those boys invite their fiancées: those girls invite all the boys they know who haven't already been invited: .... This process continues until some boy (B' say) is invited who is not already engaged.

Must that happen? Will some new unengaged boy eventually be invited? Let us represent the situation by some sets, where for ease of labelling we assume that the boys' 'names' are chosen so that they are invited in numerical order:

$\{m + 1\}$ invites $\{1', \ldots, i'\}$ invites $\{1, \ldots, i\}$ invites $\{(i + 1)', \ldots, j'\}$
invites ... invites $\{(k + 1)', \ldots, l'\}$ invites $\{k + 1, \ldots, l\}$ invites $\{\ldots B' \ldots\}$.

How do we know that this process continues until some new unengaged boy is invited? If no new unengaged boys have been invited then at a typical stage in the process, when we have a set of newly invited girls $\{k + 1, \ldots, l\}$, whom will they invite next? So far the girls $m + 1$ and $1, 2, \ldots, l$ and the boys $1', 2', \ldots, l'$ have been invited (and that includes all the boys known to girls $1, 2, \ldots$ and k), but overall that's fewer boys than girls. So, by the given condition of girls knowing at least as many boys, there still remains to be invited a boy known to one of the girls there (and hence to one of $k + 1, \ldots, l$). Therefore the process will continue. There is only a finite number of engaged boys and so at some stage the process will include a boy B' who is not already engaged.

The party is now in full swing and the boy B' dances with one of the girls who invited him. Her fiancé is a bit put out so he dances with a girl who invited him: her fiancé is a bit annoyed so he dances with a girl who invited him: ... and (as in the previous example) this traces its way back through the list of sets given above, eventually leading to a boy dancing with the girl $m+1$:

$$\{m+1\}\text{–}\{\ldots z'\ldots\}\text{–}\{\ldots z\ldots\}\text{–}\ldots\text{–}\{\ldots c'\ldots\}\text{–}\{\ldots c\ldots\}\text{–}\{\ldots B'\ldots\}$$

Pairing off some engaged couples together with girl $m+1$ and boy B' into different couples

This dance is such a success that all the dancing couples break off all former commitments and get engaged. Any couples not dancing remain engaged to their previous partners. A few moments of thought shows you that the result of this process is to rearrange the engaged people dancing and girl $m+1$ and boy B' into different engaged pairs. Hence overall we have found fiancés for all the girls $1, 2, \ldots, m+1$, and the proof by induction is complete. □

That proof is not the shortest nor the most sophisticated available but it has the advantage of actually showing how the choice of husbands can be made. It also highlights the difficulty in pairing off the couples one at a time, which is what students often try to do when providing their own 'proofs' of what they see as an easy theorem. In fact the result is non-trivial with a wide range of applications. Another proof of Hall's theorem will be found in the exercises, and further proofs can be found in Leon Mirsky's *Transversal theory*, in Robin Wilson's *Introduction to graph theory*, and in the specialised book *Independence theory in combinatorics*, all listed in the bibliography. (As an indication of the depth of the theorem it is worth noting that *Transversal theory* could be subtitled 'Hall's theorem and its corollaries' and it is packed with 255 pages of results.) Transversal theory is based upon Hall's theorem in its set-theoretic form and we now investigate the 'transversal' and graph-theoretic forms of the theorem.

A *family of sets* is an ordered list of sets $\mathfrak{A} = (A_1, \ldots, A_n)$ (and for our purposes the sets will always be finite). A *transversal* (or set of *distinct representatives*) of $\mathfrak{A}$ is a set X with $|X| = n$ which can have its elements arranged in a certain order, $X = \{a_1, \ldots, a_n\}$ say, so that $a_1 \in A_1, \ldots, a_n \in A_n$: in other words the n distinct elements of X 'represent' the n sets.

Examples Let the family $\mathfrak{A} = (A_1, A_2, A_3, A_4, A_5, A_6)$ be given by

$$A_1 = \{\mathbf{1}, \mathbf{3}\},$$
$$A_2 = \{\mathbf{2}, 3\},$$
$$A_3 = \{1, 3, \mathbf{4}, \mathbf{5}\},$$
$$A_4 = \{2, 4, 6, \mathbf{7}\},$$
$$A_5 = \{1, \mathbf{5}\},$$
$$A_6 = \{\mathbf{1}, 2\}.$$

The marriage theorem

Then the family has a transversal $X = \{1, 2, 3, 4, 5, 7\}$ (or, in a more suitable order, $\{3, 2, 4, 7, 5, 1\}$): for 3 can be used to represent A_1, $2 \in A_2$, $4 \in A_3$, $7 \in A_4$, $5 \in A_5$ and $1 \in A_6$, as shown by the bold figures. However, the family $\mathfrak{A} = (A_1, A_2, A_3, A_4, A_5, A_6)$ given by

$$A_1 = \{1, 2, 3\},$$
$$A_2 = \{2, 3\},$$
$$A_3 = \{3, 5, 7\},$$
$$A_4 = \{1, 2\},$$
$$A_5 = \{1, 2, 3\},$$
$$A_6 = \{4, 5, 6\}$$

has no transversal since, for example, the sets A_1, A_2, A_4 and A_5 only have three elements between them; i.e.

$$A_1 \cup A_2 \cup A_4 \cup A_5 = \{1, 2, 3\}. \qquad \square$$

In general it is clear that, to have any hope of finding a distinct representative for each of the sets in the family $\mathfrak{A} = (A_1, \ldots, A_n)$, we certainly need the union of any r of those sets, $A_{i_1}, A_{i_2}, \ldots, A_{i_r}$ say, to contain at least r elements; i.e.

$$\left| \bigcup_{i \in \{i_1, \ldots, i_r\}} A_i \right| \geq |\{i_1, \ldots, i_r\}|$$

This can be written more succinctly as

$$\left| \bigcup_{i \in I} A_i \right| \geq |I|$$

where I is any set of indices chosen from $\{1, \ldots, n\}$.

As some of you will have spotted, this condition and the two examples above bear a close resemblance to the earlier marriage examples. Indeed if you think of set A_1 as the set of boys whom girl 1 knows, and the set A_2 as the set of boys whom girl 2 knows, etc., then finding distinct representatives of the sets is like finding husbands for each of the girls. That observation enables us to deduce the set-theoretic version of Hall's theorem:

Corollary (Hall's theorem – transversal form) *The family $\mathfrak{A} = (A_1, \ldots, A_n)$ has a transversal if and only if*

$$\left| \bigcup_{i \in I} A_i \right| \geq |I|$$

for all $I \subseteq \{1, \ldots, n\}$.

Proof Invent n 'girls' called $1, 2, \ldots, n$ and for each i let A_i be the set of 'boys' whom girl i knows. Then applying Hall's theorem to these girls and boys gives

$\mathfrak{A}$ has a transversal $\Leftrightarrow$ the girls $1, \ldots, n$ can be found husbands from the boys whom they know

$\Leftrightarrow$ any set I of the girls know between them at least $|I|$ boys

$\Leftrightarrow$ for any I the set $\bigcup_{i \in I} A_i$ contains at least $|I|$ boys

$\Leftrightarrow \left| \bigcup_{i \in I} A_i \right| \geq |I|$ for all $I \subseteq \{1, \ldots, n\}$. □

We shall now deduce a graph-theoretic version of Hall's theorem. A graph $G = (V, E)$ is *bipartite* if the set of vertices can be partitioned into two sets V_1 and V_2 so that each edge of G joins a member of V_1 to a member of V_2. When we need to name the two sets of vertices our notation for such a graph will be $G = (V_1, E, V_2)$. A *matching from V_1 to V_2* in that graph is a set of $|V_1|$ edges with no two endpoints in common. (So each vertex in V_1 will be joined by one of the edges in the matching to a different vertex in V_2.) The complete bipartite graph $K_{m,n}$ defined in chapter 2 is an example of a bipartite graph with m of the vertices forming the set V_1 and the other n vertices forming the set V_2. If $m \leq n$ then this graph will have a matching from V_1 to V_2.

Example *Some bipartite graphs and some matchings:*

A matching from V_1 to V_2 shown in bold print

A similar matching in $K_{3,4}$

No matching from V_1 to V_2 exists □

If you think of the vertices in V_1 as 'girls' and the vertices in V_2 as 'boys', with an edge from a girl to a boy meaning that the girl knows the boy, then it is easy to see that a matching from V_1 to V_2 is like a set of marriages, choosing a different husband for each of the girls. That observation enables us to deduce the graph-theoretic version of Hall's theorem without any further proof. We also state a matrix form of the theorem and leave its proof as one of the exercises.

Corollary (Hall's theorem – graph form) *Let G be the bipartite graph (V_1, E, V_2). Then G has a matching from V_1 to V_2 if and only if given any set I of vertices in V_1 there exist at least $|I|$ vertices in V_2 joined to a member of I.* □

The marriage theorem

Corollary (Hall's theorem – matrix form) Let M be an $m \times n$ matrix of 0s and 1s. Then there exists a 1 in each row of M, no two of which are in the same column, if and only if any set of rows of M (r of them say) have between them 1s in at least r columns. □

Later we shall need a generalisation of Hall's theorem which concerns the problem of finding husbands for a set of girls and, in addition, making sure that the husbands chosen include some particular boys.

Example   Girl 1 knows boys $A_1 = \{1, 2, 7\}$,

girl 2 knows boys $A_2 = \{\mathbf{5}, \mathbf{6}, 8\}$,

girl 3 knows boys $A_3 = \{1, 3, \mathbf{7}\}$,

girl 4 knows boys $A_4 = \{2, 3, 4, 7\}$,

girl 5 knows boys $A_5 = \{1, 2, 6, \mathbf{8}\}$.

It is certainly possible to find these girls a husband each from amongst the boys they know: one such set is shown in bold print. But is it possible to find the girls one husband each so that, in particular, boys 5, 6, 7, and 8 get married? No, because, for example, the girls 1, 3 and 4 know between them the boys 1, 2, 3, 4 and 7 and so those three girls can between them only marry one of the particular boys, namely 7. So that just leaves two other girls (2 and 5) to marry the three other particular boys (5, 6 and 8), which is impossible. □

As that example illustrates, to have any hope of finding husbands for a set of girls so that, in addition, a particular set P of boys are included as husbands, we certainly need the following to hold for any subset I of girls:

$$\text{number of boys in } P \text{ not known by any girl in } I \leq \text{number of girls not included in } I.$$

We shall now show that those conditions are necessary **and sufficient**. We state and prove the result in its transversal form.

Theorem Let $\mathfrak{A} = (A_1, \ldots, A_n)$ be a family of sets and let $P \subseteq A_1 \cup \cdots \cup A_n$. Then $\mathfrak{A}$ has a transversal which includes the set P if and only if

(i) $\mathfrak{A}$ has a transversal,

and

(ii) $$\left| P \setminus \left(\bigcup_{i \in I} A_i \right) \right| \leq n - |I| \text{ for all } I \subseteq \{1, \ldots, n\}.$$

Proof (⇒) Assume that we can find a transversal for $\mathfrak{A}$ as stated. Then of course (i) follows immediately. In marriage terms we are assuming that we can find the 'girls' husbands to include the set P. As we observed above, that certainly means that for any set of girls, I, the boys in P whom they don't know must have their wives outside I. Hence the number of boys in P not known to any of the girls in I is less than or equal to the number of girls not included in I. In set-theoretic terms this translates exactly to the condition given in (ii).

(⇐) Assume that the family $\mathfrak{A}$ satisfies conditions (i) and (ii) and that $A_1 \cup \cdots \cup A_n = B$, where $|B| = m$. Now invent a new family $\mathfrak{A}^*$ of m sets given by

$$\mathfrak{A}^* = (A_1, \ldots, A_n, \underbrace{B \backslash P, B \backslash P, \ldots, B \backslash P}_{m-n \text{ times}}).$$

We claim that the given conditions (i) and (ii) ensure that $\mathfrak{A}^*$ has a transversal: we must check that any r of the sets in $\mathfrak{A}^*$ contain between them at least r elements. If those r sets are only chosen from the sets $A_1, \ldots, A_n$ then it is clear from (i) that the r sets contain between them at least r elements. If on the other hand the r sets from $\mathfrak{A}^*$ include $\{A_i : i \in I\}$ from $A_1, \ldots, A_n$ and $r - |I|$ (> 0) sets from the rest, then the union of these r sets is

$$\left(\bigcup_{i \in I} A_i \right) \cup (B \backslash P) = B \backslash \left(P \backslash \left(\bigcup_{i \in I} A_i \right) \right)$$

which by (ii) contains

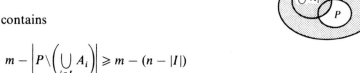

$$m - \left| P \backslash \left(\bigcup_{i \in I} A_i \right) \right| \geq m - (n - |I|)$$

elements. But $r - |I|$ is the number of copies of $B \backslash P$ chosen from $\mathfrak{A}^*$ and so it cannot exceed $m - n$. Hence

$$r - |I| \leq m - n \quad \text{and so} \quad m - (n - |I|) \geq r.$$

Therefore the r sets from $\mathfrak{A}^*$ do contain at least r elements in this case too.

It follows from the transversal form of Hall's theorem applied to $\mathfrak{A}^*$ that the family $\mathfrak{A}^*$ of m sets has a transversal as shown:

$$\mathfrak{A}^* = (A_1, \ldots, A_n, B \backslash P, B \backslash P, \ldots, B \backslash P)$$
$$\uparrow \quad\quad \uparrow \quad \uparrow \quad \uparrow \quad\quad \uparrow$$
$$a_1 \; \cdots \; a_n \; a_{n+1} \; a_{n+2} \; \cdots \; a_m$$

This transversal of m distinct representatives of $\mathfrak{A}^*$ is a subset of $A_1 \cup \cdots \cup A_n = B$, which only contains m elements in total. It follows that the transversal of $\mathfrak{A}^*$ must be B itself and in particular that it must contain the set P. But which sets in $\mathfrak{A}^*$ have the elements in P been used to represent? Certainly not any of the sets $B \backslash P$. So we have:

$$\mathfrak{A}^* = (A_1, \ldots, A_n, B\backslash P, B\backslash P, \ldots, B\backslash P)$$
$$\uparrow \quad \quad \uparrow \quad \uparrow \quad \uparrow \quad \quad \uparrow$$
$$\underbrace{a_1 \; \cdots \; a_n}_{\text{includes } P} \; a_{n+1} \; a_{n+2} \; \cdots \; a_m$$

and the set $\{a_1, \ldots, a_n\}$ is a transversal of $\mathfrak{A}$ which includes the set P. Hence the conditions (i) and (ii) do ensure the existence of a transversal of $\mathfrak{A}$ containing P, as required. $\square$

Exercises

1. In which of the following situations can husbands be found for each of the girls from amongst the boys whom they know?

   (i) Girl 1 knows boys $\{1, 2\}$,
   girl 2 knows boys $\{1, 2, 4\}$,
   girl 3 knows boys $\{1, 2, 3\}$,
   girl 4 knows boys $\{1, 3\}$,
   girl 5 knows boys $\{4, 5, 6\}$,
   girl 6 knows boys $\{1, 2, 5\}$.

   (ii) Girl 1 knows boys $\{1, 3, 5\}$,
   girl 2 knows boys $\{1, 3\}$,
   girl 3 knows boys $\{1, 5\}$,
   girl 4 knows boys $\{1, 2, 3, 4, 5\}$,
   girl 5 knows boys $\{3, 5\}$,
   girl 6 knows boys $\{2, 4, 6, 7\}$. [A]

2. (An alternative proof of Hall's theorem: do not use the theorem in your solution!)
   Let $\mathfrak{A} = (A_1, \ldots, A_n)$ be a family of sets with the Hall-type property

   $$\left| \bigcup_{i \in I} A_i \right| \geq |I|$$

   for all $I \subseteq \{1, \ldots, n\}$. Let $\mathfrak{B} = (B_1, \ldots, B_n)$ be a family of minimal subsets of $A_1, \ldots, A_n$ with the same property; i.e. $B_1 \subseteq A_1, \ldots, B_n \subseteq A_n$,

   $$\left| \bigcup_{i \in I} B_i \right| \geq |I|$$

   for all $I \subseteq \{1, \ldots, n\}$, but if any element of any B_i is removed then this Hall-type property fails.
   (i) Show that each B_i contains just one element. [H]
   (ii) Deduce that the family $\mathfrak{A}$ has a transversal.

3. Consider a set of n girls and m boys who have the property that any set of the girls (r of them, say) know at least r boys between them. In addition let B be a boy who knows at least one of the girls. By any or all of the following methods show that the girls can be found husbands with B as one of those husbands.
   (i) Adapt the 'party' proof of the marriage version of Hall's theorem.   [H]
   (ii) Invite $m - n$ new girls each of whom knows all the boys except B and apply Hall's theorem to this new situation.   [H]
   (iii) Use the last theorem of the chapter with $P = \{B\}$.   [H]

4. Show that in a group of n girls and m boys there exist some k girls for whom husbands can be found if and only if any subset of the n girls (r of them, say) know between them at least $k + r - n$ of the boys.   [H]

5. Consider a situation of girls and boys in which any set of girls know between them at least that number of boys. Assume also that there are n girls altogether and that each of them knows at least m ($\leq n$) of the boys. Show that the n marriages can be arranged in at least $m!$ ways.   [H]

6. Prove the matrix form of Hall's theorem stated as a corollary on page 31.   [H]

7. Let $\mathfrak{A} = (A_1, \ldots, A_n)$ be a family of sets in which each set contains at least d elements (where $d > 0$) and where no element is in more than d of the sets $A_1, \ldots, A_n$. Show that $\mathfrak{A}$ has a transversal.   [H]

8. Let G be the bipartite graph (V_1, E, V_2) with each vertex in V_1 of degree at least d (> 0) and each vertex in V_2 of degree d or less. Show that G has a matching from V_1 to V_2. Deduce that if each vertex of G has degree d then the edge-set E can be expressed as a disjoint union of d matchings from V_1 to V_2.

9. Let M be an $m \times n$ matrix of 0s and 1s in which each row contains at least d 1s (where $d > 0$) and no column contains more than d 1s. Show that there exist m 1s with one in each row of M and no two in the same column.
   A *permutation matrix* is a square matrix of 0s and 1s with precisely one 1 in each row and in each column. Show that a square matrix of 0s and 1s with precisely d 1s in each row and in each column is the sum of d permutation matrices.

10. An $n \times n$ matrix is *doubly stochastic* if its entries are non-negative and the entries in each of its rows and each of its columns add up to 1. Show that the $n \times n$ matrix M is doubly stochastic if and only if there exist $n \times n$ permutation matrices $M_1, \ldots, M_m$ and positive numbers $\lambda_1, \ldots, \lambda_m$ adding to 1 with
$$M = \lambda_1 M_1 + \cdots + \lambda_m M_m.$$ [H]

11. The set X is partitioned into n equal-sized sets in two ways:
$$X = X_1 \cup \cdots \cup X_n = Y_1 \cup \cdots \cup Y_n.$$

Show that there exist distinct $x_1, \ldots, x_n$ which are in different sets in both the partitions. [H]
(If you know anything about left and right cosets you might be able to see an application of this result to group theory.)

4

Three basic principles

This chapter is an illustration of three simple principles which we shall use many times in subsequent chapters.

The pigeon-hole principle

The *pigeon-hole principle* states the 'obvious' fact that if you distribute more than n items amongst n pigeon-holes then at least one pigeon-hole will have more than one item. Some of its many applications are illustrated by the following examples.

Example *Given any ten different positive integers less than 107 show that there will be two disjoint subsets of them with the same sum.*

Solution The highest numbers you could be given would be $97, 98, \ldots, 106$ which add up to 1015. So make some pigeon-holes from 0 up to 1015.

```
  □   □   □   □   ···   □     □
  0   1   2   3        1014  1015
```

Now given any set of ten positive integers less than 107 consider all its subsets (there are $2^{10} = 1024$ of them). Imagine that each subset is written on a slip of paper and that the slip is placed in the pigeon-hole corresponding to the sum of that subset. Then 1024 slips are being placed in 1016 pigeon-holes and, by the pigeon-hole principle, some pigeon-hole will have more than one slip in it. That means that two subsets of the ten integers will have the same sum. The subsets need not be disjoint but you can, of course, delete common elements from both the subsets to leave two disjoint subsets with the same sum. Alternatively if you look in all the pigeon-holes in order from 0 upwards the *first* pigeon-hole to have more than one slip in it will contain two *disjoint* subsets with the same sum. □

Example *In a group of arabs some handshaking takes place. No-one shakes their own hand and no pair of people shake hands more than once. Show that there must be two people who have shaken the same number of hands.*

Solution Let there be n arabs and let some handshaking take place as stated. Then each person will have shaken anything from 0 to $n-1$ hands. Make some pigeon-holes labelled from 0 up to $n-1$ and get each person to write his/her name on a slip of paper and put the slip in the pigeon-hole corresponding to the number of hands he/she shook.

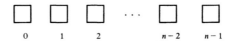

To show that two people shook the same number of hands we need to show that one of the pigeon-holes has more than one slip in it. There are n slips going into the n pigeon-holes, so we cannot be sure from the pigeon-hole principle alone that this happens. However, it can only be avoided if there is precisely one slip in each pigeon-hole. But then if Sheik Nunn's slip is in the pigeon-hole labelled 0 and Sheik All's slip is in the pigeon-hole labelled $n-1$ it means that Nunn shook hands with no-one and yet All shook hands with everyone (including Nunn), which is clearly impossible.

This result easily translates into graph-theoretic terms to say that in any graph there exist two vertices of the same degree. For if you regard the vertices of the graph as people, with two joined if they have shaken hands, then the number of hands shaken by a person is merely the degree of that vertex. □

Example *Show that, if seven points are placed on a disc of radius 1 so that no two of the points are closer than distance 1 apart, then one of the points will be at the centre of the disc and that the other six will form a regular hexagon on the circumference of the disc.*

Solution Divide the disc into seven pieces as illustrated. If seven points are placed on the disc so that no two points are closer than distance 1 apart, then no two of the points can be in the same piece. So there will be one point in each piece. That means that one point will be at the centre and the rest on the circumference. Each adjacent pair around the circumference must subtend an angle of 60° or more, and since these six angles must add up to 360° it follows that each of them is exactly 60°.

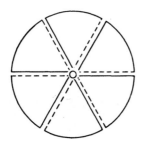 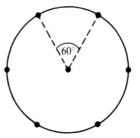

Example *Show that, given any n positive integers, there exists a subset of them whose sum is divisible by n.*

Solution Let the integers be $a_1, \ldots, a_n$ and consider the n pigeon-holes

$$\underset{0}{\square} \quad \underset{1}{\square} \quad \underset{2}{\square} \quad \cdots \quad \underset{n-2}{\square} \quad \underset{n-1}{\square}$$

Place the n subsets $\{a_1\}, \{a_1, a_2\}, \ldots, \{a_1, a_2, \ldots, a_n\}$ into the pigeon-hole corresponding to the remainder when the sum of the subset is divided by n. If any of these subsets is in the pigeon-hole 0 then that set is divisible by n. But, failing that, the n subsets will be in the $n-1$ other pigeon-holes: therefore by the pigeon-hole principle two of them, $\{a_1, a_2, \ldots, a_r\}$ and $\{a_1, a_2, \ldots, a_s\}$ say, will be in the same pigeon-hole. Since $a_1 + a_2 + \cdots + a_r$ and $a_1 + a_2 + \cdots + a_s$ give the same remainder when divided by n it follows that (if $r < s$ say) the difference $a_{r+1} + a_{r+2} + \cdots + a_s$ is divisible by n. □

Example *Prove that in a sequence of more than $(r-1)(g-1)$ different numbers there is an increasing subsequence of r terms or a decreasing subsequence of g terms (or both). (For example, any sequence of more than six numbers will have an increasing subsequence of four terms or a decreasing subsequence of three terms; e.g. the sequence* **5,** 1, **6, 3, 2,** 7, 4 *has the decreasing subsequence of three terms highlighted in bold print.)*

Solution Given a sequence $a_1, \ldots, a_n$ assign to each term a_j two 'scores', i_j and d_j, given by

$i_j =$ the number of terms in the longest increasing subsequence ending at a_j;

$d_j =$ the number of terms in the longest decreasing subsequence starting at a_j.

So, for example, the sequence 5, 1, 6, 3, 2, 7, 4 would have scores

a_j	5	1	6	3	2	7	4
i_j	1	1	2	2	2	3	3
d_j	3	1	3	2	1	2	1

In general no two terms will have the same pair of scores. For if we have

$$\cdots \quad a_j \quad \cdots \quad a_k \quad \cdots$$

then either $a_j < a_k$ and the longest increasing subsequence ending at a_j can be extended by adding on a_k (so that $i_j < i_k$) or $a_j > a_k$ and the longest decreasing subsequence starting at a_k can be preceded by a_j (so that $d_j > d_k$).

Three basic principles

Now make a grid of n^2 pigeon-holes

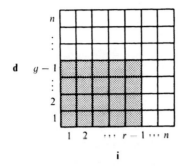

Since each i_j and d_j is between 1 and n each term of the sequence can be placed in the pigeon-hole corresponding to its two scores and, since no pair repeats itself, there will be at most one term in each pigeon-hole. In particular, if $n > (r-1)(g-1)$ then the terms will not fit in the $(r-1)(g-1)$ pigeon-holes shaded in the above picture. So some term a_j will have $i_j \geq r$ or will have $d_j \geq g$; i.e. the jth term will be either the last term of an increasing subsequence of at least r terms or the first term of a decreasing subsequence of at least g terms. □

You may well be familiar with the infinite version of that result, namely that any infinite sequence of real numbers has an infinite increasing subsequence or an infinite decreasing subsequence. There is an amusing 'Spanish hotel' proof of that result which can be found, for example, in *Yet another introduction to analysis* listed in the bibliography.

Parity

There is no one specific principle of parity which can be clearly stated, but in general terms it is used to rule out certain situations due to some sort of mismatch (such as odd/even or black/white).

Example *Show that in any graph the number of vertices of odd degree is even.*

Solution Recall from the handshaking lemma (exercise 2 on page 23) that for a graph $G = (V, E)$

$$\sum_{v \in V} \delta v = 2|\dot{E}|$$

which is even. Hence

$$\underbrace{\sum_{\substack{v \in V \\ \delta v \text{ odd}}} \delta v}_{\text{even}} + \underbrace{\sum_{\substack{v \in V \\ \delta v \text{ even}}} \delta v}_{\text{even}} = 2|E|.$$

It follows that the sum of the odd degrees is even and hence that there must be an even number of them. That is an example of the use of an odd/even 'parity' count.

□

Example *Consider an n × n board (like a chess-board) and some dominoes, each of which will cover two adjacent squares of the board. Show that the board can be covered with non-overlapping dominoes if and only if n is even. Show, however, that the board with two opposite corners removed can never be covered with non-overlapping dominoes.*

Then consider any covering of a 6 × 6 board with 18 dominoes. Show that for any such layout of dominoes it is always possible to cut the board into two rectangular pieces so that the cut does not go through a domino.

Solution It is easy to cover an even-sided board with dominoes and we leave the design of such a layout to the reader. However, if n is odd then the $\frac{1}{2}n^2$ dominoes required is not an integer!

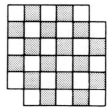

In the case of the board with two opposite corners removed we have a problem reminiscent of an age-old puzzle with a chess-board. If n is odd the covering would again require an odd half of a domino. In the case of n even imagine the board chequered, like the one illustrated. Then the two removed corners are the same colour, so the remaining board has a different number of white squares from black. Since each domino covers precisely one white and one black square the covering is again impossible. Note that in these arguments we have used both an odd/even and a black/white parity check.

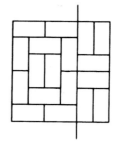

Next consider covering a 6 × 6 board with 18 dominoes. The middle figure shows one such covering and it highlights a line which cuts the board into two rectangles without cutting a domino. In general there are five possible lines in each direction to consider as possible cuts and each domino lies across just one cutting line. An odd/even parity check soon shows that in no such layout can any of those ten cuts ever go

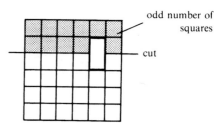

through exactly one domino (or indeed through any odd number). For, as the bottom illustration shows, that would leave an odd number of squares covered by dominoes. So how can the 18 dominoes be shared out amongst the 10 cutting lines? One cutting line must go through no dominoes, as required.

□

Three basic principles

It is worth noting that a parity check can only be used to rule out certain possibilities. For example we have just seen that a trivial odd/even count rules out the possibility of covering an odd-sided square board with dominoes. However to show that an even-sided board *can* be covered one must provide some sort of construction or proof. This will happen several times in combinatorics: a simple parity check will enable us to see when a certain layout is impossible, but to show that it is possible in all other circumstances often requires a non-trivial proof.

Example *Show that the following figure cannot be drawn with one continuous line unless one goes over a stretch of the line more than once:*

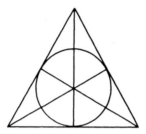

Solution Imagine any figure drawn by a continuous line without going over any stretch of the line more than once: one such figure is illustrated. Consider any point, apart from the start or the finish, where the line crosses itself. If the line has passed through that point r times then there will be $2r$ line segments leading into or out of that point. So in any figure which can be drawn in this way there will be at most two points (namely the starting-point and the finishing-point) with an odd number of line segments meeting there. In the figure given in the example there are more than two 'odd' junctions and so it cannot be drawn in this way. □

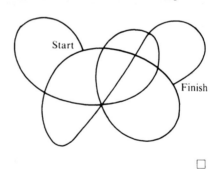

That easy parity-type observation has a far less trivial converse, namely that any connected figure with two or less 'odd' junctions *can* be drawn without taking pen from the paper and without going over any stretch of line twice. We shall study this problem in graph-theoretical terms in chapter 6.

Example *Show that in a bipartite graph each cycle uses an even number of edges.*

Solution Let the bipartite graph be $G = (V_1, E, V_2)$ (in which the vertex-set is partitioned into V_1 and V_2 and each edge joins a member of V_1 to a member of V_2). Then consider the cycle

$$v_1, v_2, v_3, \ldots, v_{n-1}, v_n, v_1$$

in G. Suppose, for instance, that $v_1 \in V_1$. Then the existence of the edge $v_1 v_2$ ensures that $v_2 \in V_2$: similarly $v_3 \in V_1$ etc. So we have

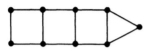

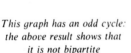

which clearly implies that n is even. □

Again the converse of that result turns out to be true, namely that if a graph has no cycles using an odd number of edges then it must be bipartite. However the proof is rather more substantial, as we shall see after illustrating the result with some examples.

Example

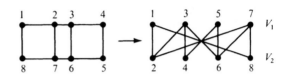

This graph has an odd cycle: the above result shows that it is not bipartite

Here there is no odd cycle: it is bipartite

Theorem *A graph is bipartite if and only if it has no cycle which uses an odd number of edges.*

Proof ($\Rightarrow$) If a graph is bipartite then, as we observed above, it cannot have an 'odd' cycle.

($\Leftarrow$) Now assume that the graph $G = (V, E)$ has no odd cycle. Assume also that G is connected (for if not we could apply this proof to each of its components, and if each component is bipartite then so is the original graph). Let $v_0 \in V$ and define $V_1, V_2 \subseteq V$ by

$$V_1 = \{v \in V: \text{ there is a path in } G \text{ from } v_0 \text{ to } v \text{ which uses an even number of edges}\},$$

$$V_2 = \{v \in V: \text{ there is a path in } G \text{ from } v_0 \text{ to } v \text{ which uses an odd number of edges}\}.$$

Since G is connected it is clear that $V = V_1 \cup V_2$, but can we be sure that $V_1 \cap V_2 = \emptyset$ and that each edge joins a member of V_1 to a member of V_2?

To show that $V_1 \cap V_2 = \emptyset$ we shall assume that $v \in V_1 \cap V_2$ and deduce a contradiction. Since $v \in V_1$ there must exist a path from v_0 to v which uses a set of edges E_1, say, where $|E_1|$ is even. Similarly, as $v \in V_2$, there must exist a path from v_0 to v which uses a set of edges E_2, say, where $|E_2|$ is odd. Let E^* be the 'symmetric difference' of E_1 and E_2; i.e.

$$E^* = (E_1 \cup E_2) \setminus (E_1 \cap E_2),$$

Three basic principles

and let G^* be the graph (V, E^*). A few moments of thought will show you that the degree of a vertex in G^* equals its degree in (V, E_1) plus its degree in (V, E_2) minus twice its degree in $(V, E_1 \cap E_2)$ and that this is always even. Therefore, by the result of exercise 3 on page 23, there exists a collection of cycles in G^* (and hence in G) with no edges in common which between them use all the edges in E^*. Since, by assumption, each such cycle is even it follows that $|E^*|$ is even.

We now wish to count the number of edges in E^*. If we simply add the number in E_1 to the number in E_2, then each member of $E_1 \cap E_2$ will have been counted twice whereas they should not have been counted at all. Hence it is clear that

$$|E^*| = |E_1| + |E_2| - 2|E_1 \cap E_2|$$
$$\text{even} \quad \text{even} \quad \text{odd} \quad \quad \text{even}$$

and we have derived a contradiction. It follows that $V_1 \cap V_2$ is empty, as claimed.

Finally it follows easily that each edge of G joins a member of V_1 to a member of V_2. For if $v_1 \in V_1$ and $v_1 v_2 \in E$ then there is a path P from v_0 to v_1 using the lowest possible even number of edges, and it is not hard to see that there is then an 'odd' path from v_0 to v_2. For if P does not end with the edge $v_2 v_1$ then the path consisting of P followed by the edge $v_1 v_2$ gives an odd path from v_0 to v_2. Alternatively, if the path P does end with the edge $v_2 v_1$ then P with the edges $v_2 v_1$ deleted gives an odd path from v_0 to v_2. In both cases the odd path from v_0 to v_2 shows that $v_2 \in V_2$ and it follows that G is the bipartite graph (V_1, E, V_2). □

The inclusion/exclusion principle

We have already seen (as in the last proof) that to count the members of the set $X \cup Y$ we must add those in X to those in Y but then deduct those which we have counted twice, namely those in both X and Y; i.e.

$$|X \cup Y| = |X| + |Y| - |X \cap Y|.$$

The 'inclusion/exclusion principle' generalises that idea and it can be stated in terms of sets or, more conveniently for us, in terms of objects having certain properties.

Example *In a club there are 10 people who play tennis and 15 who play squash: 6 of them play both. How many people play at least one of the sports?*

Solution As in the set-theoretic result quoted above, the required number here is $10 + 15 - 6 = 19$. □

Example *In a club there are 10 people who play tennis, 15 who play squash and 12 who play badminton. Of these, 5 play tennis and squash, 4 play tennis and badminton and 3 play squash and badminton: and of these just 2 people play all three sports. How many people play at least one of the three sports?*

Solution We start by adding 10, 15 and 12. But those who play two sports will have been counted twice so we must 'exclude' the 5, 4 and 3 to give

$$10 + 15 + 12 - 5 - 4 - 3.$$

But then the players of all three sports will have been included three times in the 10, 15 and 12 and then excluded three times in the 5, 4 and 3 so they will not have been counted at all. So we must include again all such people to give

$$10 + 15 + 12 - 5 - 4 - 3 + 2 = 27$$

people playing at least one of the sports. □

In those examples you can begin to see the principle of inclusion/exclusion at work.

Theorem (The inclusion/exclusion principle) *Given a finite set of objects which may or may not have any of the properties $1, 2, \ldots, n$, let $N(i_1, \ldots, i_r)$ be the number of those objects which have at least the r properties $i_1, \ldots, i_r$. Then the number of objects in the set having at least one of the properties is*

$$N(1) + N(2) + \cdots + N(n)$$
$$- N(1, 2) - N(1, 3) - \cdots - N(n-1, n)$$
$$+ N(1, 2, 3) + N(1, 2, 4) + \cdots + N(n-2, n-1, n)$$
$$- \cdots$$
$$\vdots$$
$$+ (-1)^{n-1} N(1, 2, \ldots, n).$$

Proof It is clear that if an object in the set has none of the properties then it will not contribute to the grand total stated in the theorem. We shall now show that each object in the set which satisfies at least one of the properties contributes exactly one to the grand total stated: it will then follow that the total given equals the number of such objects. Assume then that we have an object satisfying at least one of the properties and assume that it satisfies precisely r of them. (If you want to be more specific you can assume that it satisfies precisely the properties $1, 2, \ldots, r$.) Then what does this object contribute to the following grand total?

$$N(1) + N(2) + \cdots + N(n) \qquad \text{row 1}$$
$$- N(1, 2) - N(1, 3) - \cdots - N(n-1, n) \qquad \text{row 2}$$
$$+ N(1, 2, 3) + N(1, 2, 4) + \cdots + N(n-2, n-1, n) \qquad \text{row 3}$$
$$- \cdots$$
$$\vdots$$
$$+ (-1)^{r-1} N(1, 2, \ldots, r) + (-1)^{r-1} N(1, 2, \ldots, r-1, r+1) + \cdots \qquad \text{row } r$$
$$\vdots$$
$$+ (-1)^{n-1} N(1, 2, \ldots, n). \qquad \text{row } n$$

Three basic principles

In row 1 the object in question contributes $+1$ on r occasions, in row 2 it contributes -1 on $\binom{r}{2}$ occasions, in row 3 it contributes $+1$ on $\binom{r}{3}$ occasions, ... and in row r it contributes $(-1)^{r-1}$ on $\binom{r}{r}$ occasions: thereafter it contributes nothing. So its total contribution is

$$\binom{r}{1} - \binom{r}{2} + \binom{r}{3} - \cdots + (-1)^{r-1}\binom{r}{r}$$

$$= 1 - \underbrace{\left[\binom{r}{0} + \binom{r}{1}(-1) + \binom{r}{2}(-1)^2 + \cdots + \binom{r}{r}(-1)^r\right]}_{\text{the binomial expansion of } [1 + (-1)]^r \text{ which is clearly zero}} = 1$$

It follows that an object with at least one of the properties does contribute exactly one to the given total, and the principle of inclusion/exclusion is proved. □

Example *How many numbers from 2 to 1000 are perfect squares, perfect cubes or any higher power?*

Solution Consider the set of objects $\{2, \ldots, 1000\}$ and let a member of this set have 'property i' if it equals the ith power of some integer. Since $2^{10} > 1000$ there are no tenth powers in the set and the only properties which really concern this example are properties $2, 3, \ldots, 9$. Then, using the notation of the theorem, the number of objects with at least one of these properties is

$$N(2) + N(3) + \cdots + N(9)$$
$$- N(2, 3) - N(2, 4) - \cdots - N(8, 9)$$
$$+ N(2, 3, 4) + N(2, 3, 5) + \cdots + N(7, 8, 9)$$
$$- \cdots$$
$$\vdots$$
$$- N(2, 3, \ldots, 9).$$

Each of these numbers is easy to calculate. For example

$$N(2) = [\sqrt[2]{1000}] - 1 = 30 \qquad N(3) = [\sqrt[3]{1000}] - 1 = 9 \qquad \cdots$$
$$N(2, 3) = N(6) = [\sqrt[6]{1000}] - 1 = 2 \qquad N(2, 4) = N(4) = 4 \qquad \cdots$$
$$N(2, 3, 4) = N(12) = 0 \qquad N(2, 3, 6) = N(6) = 2 \qquad \cdots$$

(where $[x]$ denotes the 'integer part' of x). Continuing in this way gives the number of objects with at least one of the properties as

$$30 + 9 + 4 + 2 + 2 + 1 + 1 + 1 - 2 - 4 - 2 - 1 - 2 - 1 - 1 + 2 + 1 = 40. \quad \Box$$

Example (Euler's function) *Let m be a positive integer whose distinct prime factors are $p_1, \ldots, p_n$. Show that the number of integers between 1 and m which are relatively prime to m (i.e. with no factor greater than 1 in common with m) equals*

$$m\left(1 - \frac{1}{p_1}\right)\left(1 - \frac{1}{p_2}\right) \cdots \left(1 - \frac{1}{p_n}\right).$$

(This expression is usually denoted by $\phi(m)$ and it defines Euler's function *in number theory.)*

Solution Let the set of objects in question be $\{1, 2, \ldots, m\}$ and for $1 \leq i \leq n$ let 'property i' be that a number is divisible by p_i. Then the integers in that set which are relatively prime to m are precisely those which have none of the properties $1, \ldots, n$. So we must subtract the number of integers with at least one of the properties from m to give (with the usual notation)

$$\begin{aligned}
&m \\
&- N(1) - N(2) - \cdots + N(n) \\
&+ N(1, 2) + N(1, 3) + \cdots + N(n-1, n) \\
&- N(1, 2, 3) + N(1, 2, 4) - \cdots - N(n-2, n-1, n) \\
&+ \cdots \\
&\vdots \\
&+ (-1)^n N(1, 2, \ldots, n).
\end{aligned}$$

Now $N(i)$ is the number of integers in $\{1, \ldots, m\}$ which are divisible by p_i and that is precisely m/p_i. Similarly $N(i, j) = m/p_i p_j$, etc. Hence the required number of integers relatively prime to m is given by

$$\begin{aligned}
&m \\
&- \frac{m}{p_1} - \frac{m}{p_2} - \cdots \\
&+ \frac{m}{p_1 p_2} + \frac{m}{p_1 p_2} + \cdots \\
&- \frac{m}{p_1 p_2 p_3} - \frac{m}{p_1 p_2 p_3} - \cdots \\
&+ \cdots \\
&\vdots \\
&+ (-1)^n \frac{m}{p_1 p_2 \cdots p_n}.
\end{aligned}$$

Three basic principles

It is straightforward to check that this agrees with the given factorised version of $\phi(m)$. □

Example *How many of the permutations of $\{1, 2, \ldots, n\}$ have the property that $1 \not\to 1, 2 \not\to 2, \ldots,$ and $n \not\to n$? Such a permutation, in which each entry 'moves', is called a derangement of $\{1, 2, \ldots, n\}$.*

Solution Let the set of objects in question be all $n!$ permutations of $\{1, 2, \ldots, n\}$. For $1 \leq i \leq n$ let 'property i' of a permutation be that $i \to i$. Then the number of derangements of $\{1, \ldots, n\}$ is the number of permutations with none of these properties. By the inclusion/exclusion principle this number is given by

$$n!$$
$$- N(1) - N(2) - \cdots + N(n)$$
$$+ N(1, 2) + N(1, 3) + \cdots + N(n-1, n)$$
$$- N(1, 2, 3) + N(1, 2, 4) - \cdots - N(n-2, n-1, n)$$
$$\vdots$$
$$+ (-1)^n N(1, 2, \ldots, n).$$

But $N(1, 2, \ldots, r)$, for example, is the number of permutations of $\{1, \ldots, n\}$ in which $1 \to 1, 2 \to 2, \ldots, r \to r$ and the remaining $n - r$ entries are freely permuted: hence this number is $(n - r)!$. The same is true for any of the choices of r properties from the n and it follows that the expression above for the number of derangements becomes

$$n! - \binom{n}{1}(n-1)! + \binom{n}{2}(n-2)! - \cdots + (-1)^{n-1}\binom{n}{n-1}1! + (-1)^n\binom{n}{n}0!$$

It is easy to use the factorial versions of the binomial coefficients to reduce this expression to

$$n!\left(\frac{1}{2!} - \frac{1}{3!} + \frac{1}{4!} - \cdots + (-1)^n \frac{1}{n!}\right).$$
□

We shall use the principle of inclusion/exclusion again in subsequent chapters and, in particular, in our work on 'rook polynomials' in chapter 12.

Exercises

1. Show that
   (i) there exist two people in London who own the same number of books,
   (ii) there exist two people in the world with the same number of hairs.

2. Suppose that you are given $n + 1$ different positive integers less than or equal to $2n$. Show that

(i) there exists a pair of them which adds up to $2n + 1$, [H]
(ii) there must exist two which are relatively prime (i.e. that have no factors larger than 1 in common), [H]
(iii) there exists one which is a multiple of another. [H]

3. Prove that in any $n + 1$ integers there will be a pair which differs by a multiple of n. [H]

4. Let T be an equilateral triangle of side 1. Show that
   (i) if five points are placed in T two of them must be distance $\frac{1}{2}$ or less apart, [H]
   (ii) it is impossible to cover T with three circles each of diameter less than $1/\sqrt{3}$. [H]

5. Prove that, given any positive integer n, some multiple of it must be of the form $99\cdots900\cdots0$. [H]

6. Every day I put either 1p or 2p into a piggy-bank and the total is m pence after n days. Show that for any integer k with $0 \leq k \leq 2n - m$ there will have been a period of consecutive days during which the total amount put into the piggy-bank was exactly k pence. [H]

7. Show that the sum of two odd perfect squares cannot be a perfect square.

8. An $n \times n$ black and white chequered board has two of its squares removed. Show that the remaining board can be covered with non-overlapping dominoes if and only if n is even and the two removed squared are of different colours. [H]

9. On an $n \times n$ board there are n^2 chess pieces, one on each square. I wish to move each piece to an adjacent square in the same row or column (referred to in chess as the rank and file) so that after all the n^2 pieces have moved there is still one piece on each square. Show that this is possible if and only if n is even.

10. I have $\frac{1}{2}(n^3 - (n-2)^3)$ bricks each 1 cm × 1 cm × 2 cm and I wish to glue them together to make the outer shell of an $n \times n \times n$ cube (i.e. a cube of side n cm with a cube of side $(n-2)$ cm cut out from its centre). Show that this can be done if and only if n is even. [H]

11. (i) In a graph a *Hamiltonian cycle* is a cycle which uses each vertex of the graph. (We shall study these in detail in chapter 6.) Show that a bipartite graph $G = (V_1, E, V_2)$ which has a Hamiltonian cycle must have $|V_1| = |V_2|$. [H]
    (ii) In chess a knight can move in any direction across the diagonal of a 2 × 3 rectangle, as shown. Given an $n \times n$ board, where $n > 1$ is odd, show that it is impossible for a knight to start at one square and move around the board visiting each square once and then ending back where it started. [H]
    Keen readers might like to find such a route in the case of a traditional 8 × 8 chess-board. [A]

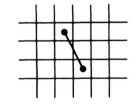

12. (i) How many integers between 1 and 10 000 are divisible by at least one of 2, 3 and 5? [A]
    (ii) How many integers between 1 and 10 000 are divisible by at least one of 2, 3, 5 and 7? [A]
    Deduce that there are at most 2288 primes less than 10 000.
    (We could reduce this number by gradually excluding multiples of the lowest remaining prime, imitating the famous *Sieve of Eratosthenes*.)

13. A poker hand consists of 5 cards from the normal pack of 52. By each of the following methods count the number of different poker hands which contain at least one ace, at least one king, at least one queen and at least one jack:
    (i) inclusion/exclusion;
    (ii) counting the various types of hands, such as AAKQJ. [A]

14. Some n letters have been typed and so have their envelopes. If the letters are placed with one in each envelope in a random fashion what is the probability that no letter is in the correct envelope? Show that this probability tends to $1/e$ as n tends to infinity. [A]

15. (i) How many functions are there from $\{1, \ldots, m\}$ to $\{1, \ldots, n\}$? [A]
    (ii) An *injection* from $\{1, \ldots, m\}$ to $\{1, \ldots, n\}$ is a function f such that if $i \neq j$ in $\{1, \ldots, m\}$ then $f(i) \neq f(j)$ in $\{1, \ldots, n\}$. How many of the functions from $\{1, \ldots, m\}$ to $\{1, \ldots, n\}$ are injections? [A]
    (iii) A *surjection* from $\{1, \ldots, m\}$ to $\{1, \ldots, n\}$ is a function f such that for each $i \in \{1, \ldots, n\}$ there exists $j \in \{1, \ldots, m\}$ with $f(j) = i$. Use the inclusion/exclusion principle to show that the number of surjections from $\{1, \ldots, m\}$ to $\{1, \ldots, n\}$ is given by

$$n^m - \binom{n}{1}(n-1)^m + \binom{n}{2}(n-2)^m - \cdots + (-1)^{n-1}\binom{n}{n-1}1^m.$$ [H]

5
Latin squares

A 'Latin square' is a square array or matrix in which each row and each column consists of the same set of entries without repetition.

Example *From the world of abstract algebra we note that the main body of the multiplication table of a finite group forms a Latin square. For example the* Klein *group has multiplication table*

	a	b	c	d
a	a	b	c	d
b	b	a	d	c
c	c	d	a	b
d	d	c	b	a

a Latin square

Example *From probability theory the particular 'doubly stochastic' matrix*

$$\begin{pmatrix} 0 & \frac{1}{4} & \frac{3}{4} \\ \frac{1}{4} & \frac{3}{4} & 0 \\ \frac{3}{4} & 0 & \frac{1}{4} \end{pmatrix}$$

(in which each row and column adds to 1) is a Latin square. □

We shall generally restrict attention to Latin squares and rectangles in which the entries are positive integers. A $p \times q$ *Latin rectangle* (with entries in $\{1, 2, \ldots, n\}$) is a $p \times q$ matrix with entries chosen from $\{1, 2, \ldots, n\}$ and with no repeated entry in any row or column. In the cases when $p = q = n$ it is called a *Latin square*: in that case each row and each column consists precisely of the n numbers $1, 2, \ldots$ and n.

Latin squares

Examples

$$\begin{pmatrix} 1 & 5 & 2 \\ 4 & 2 & 1 \end{pmatrix} \qquad \begin{pmatrix} 3 & 1 & 2 \\ 1 & 2 & 3 \\ 2 & 3 & 1 \end{pmatrix} \qquad \begin{pmatrix} 1 & 2 & 4 & 3 \\ 3 & 4 & 2 & 1 \\ 4 & 3 & 1 & 2 \\ 2 & 1 & 3 & 4 \end{pmatrix}$$

A 2×3 Latin rectangle with entries in $\{1, 2, 3, 4, 5\}$ $\quad$ A 3×3 Latin square $\quad$ A 4×4 Latin square $\quad \square$

In this chapter we shall consider two main questions concerning Latin rectangles and squares: the first is whether a Latin rectangle is necessarily part of a Latin square.

Example *Is the Latin rectangle*

$$\begin{pmatrix} 1 & 2 & 3 & 4 \\ 4 & 3 & 1 & 2 \end{pmatrix}$$

part of a 4×4 Latin square? Equivalently, can two further rows be added to make such a square? Similarly, is the Latin rectangle

$$\begin{pmatrix} 1 & 3 & 4 & 5 \\ 3 & 5 & 1 & 2 \\ 5 & 1 & 3 & 4 \end{pmatrix}$$

part of a 5×5 Latin square? Equivalently, can a further column and then two further rows be added to make such a square?

Solution It is easy to see that the first Latin rectangle can be extended to a Latin square, one such example being

$$\begin{pmatrix} 1 & 2 & 3 & 4 \\ 4 & 3 & 1 & 2 \\ 2 & 1 & 4 & 3 \\ 3 & 4 & 2 & 1 \end{pmatrix}$$

Indeed we shall soon prove that any $p \times n$ Latin rectangle with entries in $\{1, \ldots, n\}$ can be extended to an $n \times n$ Latin square.

However, the second example cannot be extended to a 5×5 Latin square because the additional column in that extension would have to include a '2' in two places and so could not be Latin: in some sense the '2' did not occur enough times in the given rectangle. Later in this chapter we shall obtain conditions on the number of occurrences of the entries in a Latin rectangle which are necessary and sufficient to ensure that it can be extended to a Latin square. $\quad \square$

Our results concerning the extensions of Latin rectangles will be proved using the transversal form of Hall's theorem from chapter 3. We now give one example to show how the addition of an extra row to a Latin rectangle translates easily to a problem in transversals.

Example *To add an extra row to the Latin rectangle*

$$\begin{pmatrix} 1 & 2 & 4 & 5 & 3 \\ 5 & 1 & 2 & 3 & 4 \end{pmatrix}$$

to create a 3×5 *Latin rectangle with entries in* $\{1, 2, 3, 4, 5\}$ *is equivalent to finding distinct representatives of the sets shown below (note that each set has cardinality 3 and that each number 1–5 is in exactly three of the sets):*

$$\begin{pmatrix} 1 & 2 & 4 & 5 & 3 \\ 5 & 1 & 2 & 3 & 4 \end{pmatrix}$$
$$\{\mathbf{2}, 3, 4\} \quad \{3, \mathbf{4}, 5\} \quad \{\mathbf{1}, 3, 5\} \quad \{1, \mathbf{2}, 4\} \quad \{1, 2, \mathbf{5}\}$$

One such collection is shown in bold print: those representatives could be used as the entries of the next row. We could then continue in a similar way to extend the 3×5 *Latin rectangle to a* 4×5 *Latin rectangle and finally to a* 5×5 *Latin square.* □

Theorem *Any* $p \times n$ *Latin rectangle with entries in* $\{1, \ldots, n\}$ *can be extended to an* $n \times n$ *Latin square.*

Proof Assume that L is a $p \times n$ Latin rectangle with $p < n$, and for $1 \leq i \leq n$ let the set A_i be given by

$A_i = \{$those numbers from $1, 2, \ldots, n$ not used in the ith column of $L\}$

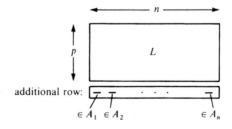

Then, as we saw in the above example, finding an additional row for L to extend it to a $(p + 1) \times n$ Latin rectangle (with entries in $\{1, \ldots, n\}$) is equivalent to finding distinct representatives of the family $\mathfrak{A} = (A_1, \ldots, A_n)$. To show that such representatives exist, or that this family has a transversal, we use Hall's theorem (page 29) and show that any r of these sets have between them at least r elements.

Note before proceeding that (as in the previous example)

$$|A_1| = |A_2| = \cdots = |A_n| = n - p$$

Latin squares

because A_i consists of all n numbers except for those p already occurring in the ith column. Note also (again as in the example) that each j with $1 \leq j \leq n$ lies in precisely $n - p$ of the sets in $\mathfrak{A}$. For j occurs in precisely p places in L and so it occurs in just p different columns: j is therefore missing from $n - p$ columns and hence it occurs in just those $n - p$ sets.

We have already seen in exercise 7 on page 34 that such a symmetrical family must have a transversal, but we again prove this directly from Hall's theorem. Consider any r of the sets from $A_1, \ldots, A_n$: we must show that these r sets contain between them at least r elements. So list the total entries of all these r sets (including repeats). There will be $r(n - p)$ entries in this list. But, by the above comments, no number j can be there more than $n - p$ times and so this list must clearly contain at least r *different* numbers. It follows that any r sets in $\mathfrak{A}$ contain between them at least r different numbers and, by Hall's theorem, $\mathfrak{A}$ has a transversal.

As commented above, any transversal of $\mathfrak{A}$ can be used to form an additional row for L in an obvious way to give a $(p + 1) \times n$ Latin rectangle. We can then start all over again and keep repeating the process until an $n \times n$ Latin square is reached. □

The theorem showed that we can always extend $p \times \mathbf{n}$ Latin rectangles to $n \times n$ Latin squares but we now turn attention to $p \times q$ rectangles: as we saw earlier, extension is not always possible.

Example *Can the Latin rectangle*

$$\begin{pmatrix} 6 & 1 & 2 & 3 \\ 5 & 6 & 3 & 1 \\ 1 & 3 & 6 & 2 \\ 3 & 2 & 4 & 6 \end{pmatrix}$$

be extended to a 6×6 Latin square?

Solution No. One way to see this is to note that in any extension to a 6×6 Latin square as shown we would need three 5s in the box, but they will not fit because there are only two columns for them.

$$\begin{pmatrix} 6 & 1 & 2 & 3 & \_ & \_ \\ 5 & 6 & 3 & 1 & \_ & \_ \\ 1 & 3 & 6 & 2 & \_ & \_ \\ 3 & 2 & 4 & 6 & \_ & \_ \\ \_ & \_ & \_ & \_ & \_ & \_ \\ \_ & \_ & \_ & \_ & \_ & \_ \end{pmatrix}$$

□

In general to have any hope of extending a $p \times q$ Latin rectangle (with entries in $\{1, \ldots, n\}$) to an $n \times n$ Latin square there must be restrictions on the number of occurrences of the entries in L. Imagine that L can be extended, as illustrated below, and that the number i occurs precisely $L(i)$ times in L.

For i to occur once in each row of the eventual square we need it to occur p times in the first p rows, and hence $p - L(i)$ times in the top right-hand area. But there are only $n - q$ columns there. So to have any hope of extending L to an $n \times n$ Latin square we certainly need $p - L(i) \leqslant n - q$ or

$$L(i) \geqslant p + q - n \text{ for each } i \in \{1, \ldots, n\}.$$

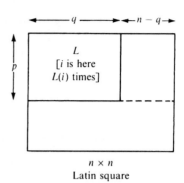

Remarkably these conditions turn out to be both necessary and sufficient to ensure that L can be extended to an $n \times n$ Latin square. Again we shall use our work on transversals to prove this result, but our choice of transversal must be made with care, as the following example illustrates.

Example *Use the notation of transversals to show how*

$$\begin{pmatrix} 1 & 3 & 4 \\ 4 & 1 & 5 \end{pmatrix}$$

can be extended to a 5×5 Latin square.

Solution Here the given L is 2×3 and so $p = 2$, $q = 3$ and $n = 5$. Thus $p + q - n$ is 0 and it is clear that $L(i) \geqslant p + q - n$ for each i. We shall start by extending L to a 2×4 Latin rectangle by the addition of an extra column:

$$\begin{pmatrix} 1 & 3 & 4 & - \\ 4 & 1 & 5 & - \end{pmatrix} \begin{matrix} \nearrow \in \{2, 5\} \\ \searrow \in \{2, 3\} \end{matrix}$$

This can be done in any one of three ways; in each case we can then try to extend to a 2×5 Latin rectangle:

Latin squares

$$\begin{pmatrix} 1 & 3 & 4 & 2 \\ 4 & 1 & 5 & 3 \end{pmatrix} \quad \begin{pmatrix} 1 & 3 & 4 & 5 \\ 4 & 1 & 5 & 2 \end{pmatrix} \quad \begin{pmatrix} 1 & 3 & 4 & 5 \\ 4 & 1 & 5 & 3 \end{pmatrix}$$

$$\downarrow \qquad\qquad \downarrow \qquad\qquad \downarrow$$

$$\begin{pmatrix} 1 & 3 & 4 & 2 & 5 \\ 4 & 1 & 5 & 3 & 2 \end{pmatrix} \quad \begin{pmatrix} 1 & 3 & 4 & 5 & 2 \\ 4 & 1 & 5 & 2 & 3 \end{pmatrix} \qquad \text{Impossible}$$

In these two cases we now have a 2 × 5 Latin rectangle and, by the previous theorem, these can be extended to 5 × 5 Latin squares.

Here the 2 × 4 Latin rectangle could not be extended because $p = 2$, $q = 4$ and $p + q - n = 1$. But $L(2) = 0$ and so $L(2) < p + q - n$. □

Assume that we are given a $p \times q$ Latin rectangle L with entries in $\{1, \ldots, n\}$, with $L(i) \geq p + q - n$ for $1 \leq i \leq n$, and that we wish to extend L to an $n \times n$ Latin square. The moral of the above example is that when we add a column to L to give a $p \times (q + 1)$ Latin rectangle L' we must do it in such a way that the process can then be repeated. So we require that $L'(i)$, the number of occurrences of i in L', must satisfy

$$L'(i) \geq p + (q + 1) - n.$$

Therefore, if for some i we have $L(i) = p + q - n$, then we must ensure that i is included in the new column. So let

$$P = \{i\colon 1 \leq i \leq n \text{ and } L(i) = p + q - n\}.$$

Then for the process to continue the extra column must include the set P, for in that event each i will occur in the new rectangle at least $p + (q + 1) - n$ times.

Example *Use the above techniques to extend*

$$\begin{pmatrix} 5 & 6 & 1 \\ 6 & 5 & 2 \\ 1 & 2 & 3 \end{pmatrix}$$

to a 6 × 6 Latin square.

Solution We shall follow the process through in full because it shows the significance of the set P and it will help us to understand the proof of the next theorem.

$$\begin{pmatrix} 5 & 6 & 1 \\ 6 & 5 & 2 \\ 1 & 2 & 3 \end{pmatrix} \begin{array}{l} \longrightarrow \in \{2, 3, 4\} \\ \longrightarrow \in \{1, 3, 4\} \\ \longrightarrow \in \{4, 5, 6\} \end{array}$$

$$p + q - n = 3 + 3 - 6 = 0$$

The new column must include the set

$$P = \{i\colon L(i) = 0\} = \{4\}.$$

One such transversal is shown in bold print and the process can continue:

$$\begin{pmatrix} 5 & 6 & 1 & 4 \\ 6 & 5 & 2 & 1 \\ 1 & 2 & 3 & 5 \end{pmatrix} \begin{matrix} \in \{2, 3\} \\ \in \{3, 4\} \\ \in \{4, 6\} \end{matrix}$$

$$p + q - n = 3 + 4 - 6 = 1.$$

The new column must include the set

$$P = \{i: L(i) = 1\} = \{3, 4\}.$$

One such transversal is shown in bold print and the process can continue:

$$\begin{pmatrix} 5 & 6 & 1 & 4 & 3 \\ 6 & 5 & 2 & 1 & 4 \\ 1 & 2 & 3 & 5 & 6 \end{pmatrix} \begin{matrix} \in \{2\} \\ \in \{3\} \\ \in \{4\} \end{matrix}$$

$$p + q - n = 3 + 5 - 6 = 2.$$

The new column must include the set

$$P = \{i: L(i) = 2\} = \{2, 3, 4\}.$$

One such transversal (!) is shown in bold print and it gives the 3×6 Latin rectangle

$$\begin{pmatrix} 5 & 6 & 1 & 4 & 3 & 2 \\ 6 & 5 & 2 & 1 & 4 & 3 \\ 1 & 2 & 3 & 5 & 6 & 4 \end{pmatrix}.$$

By the previous theorem this can now be extended to a 6×6 Latin square, one such being

$$\begin{pmatrix} 5 & 6 & 1 & 4 & 3 & 2 \\ 6 & 5 & 2 & 1 & 4 & 3 \\ 1 & 2 & 3 & 5 & 6 & 4 \\ 2 & 3 & 4 & 6 & 1 & 5 \\ 4 & 1 & 5 & 3 & 2 & 6 \\ 3 & 4 & 6 & 2 & 5 & 1 \end{pmatrix}. \qquad \square$$

We are now almost ready to prove our result about the extensions of $p \times q$ Latin rectangles, but we shall also need the following lemma:

Lemma *Let L be a $p \times q$ Latin rectangle with entries in $\{1, \ldots, n\}$ and let r and m be integers with $0 \leqslant r \leqslant m < p$. Then the number of members of $\{1, \ldots, n\}$ which occur exactly m times in L and which occur in all of the first r rows of L cannot exceed*

$$\frac{(n-q)(p-r)}{p-m}.$$

Latin squares

Proof Let t members of $\{1, \ldots, n\}$ occur exactly m times in L and occur in all the first r rows of L. Then those t numbers each occur exactly $m - r$ times in the lower shaded region of L. The other $n - t$ numbers in $\{1, \ldots, n\}$ each occur at most $p - r$ times in that same shaded region. So counting the total occurrences in that shaded region gives

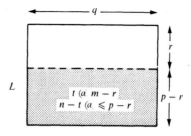

$$(p - r)q \leq t(m - r) + (n - t)(p - r)$$

which reduces to

$$t(p - m) \leq (n - q)(p - r)$$

as required. □

Theorem *The $p \times q$ Latin rectangle L with entries in $\{1, \ldots, n\}$ can be extended to an $n \times n$ Latin square if and only if $L(i)$, the number of occurrences of i in L, satisfies*

$$L(i) \geq p + q - n$$

for each i with $1 \leq i \leq n$.

Proof ($\Rightarrow$) Assume first that L can be extended to an $n \times n$ Latin square as shown:

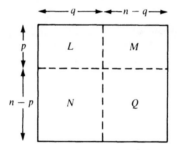

Then i occurs $L(i)$ times in L, and p times in L and M together. So i occurs $p - L(i)$ times in M. But i occurs $n - q$ times in M and Q together. Hence

$$p - L(i) \leq n - q \quad \text{and so} \quad L(i) \geq p + q - n.$$

($\Leftarrow$) Conversely assume that $L(i) \geq p + q - n$ for each i and assume that $q < n$. We shall show that L can be extended to a $p \times (q + 1)$ Latin rectangle in such a way that the process can be continued until a $p \times n$ Latin rectangle is reached. (Then the previous theorem ensures that such a rectangle can be extended to an $n \times n$ Latin square.)

For $1 \leq i \leq p$ let the set A_i be given by

$$A_i = \{\text{those numbers from } 1, 2, \ldots, n \text{ not used in the } i\text{th row of } L\}$$

and, as in the examples, let the set P be given by

$$P = \{i \colon 1 \leq i \leq n \text{ and } L(i) = p + q - n\}.$$

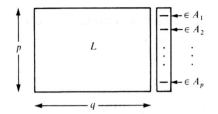

Then we shall show that $\mathfrak{A} = (A_1, \ldots, A_p)$ has a transversal which includes the set P. (We shall then use this transversal to give an extra column to extend L to a $p \times (q+1)$ Latin rectangle L'.)

To show that $\mathfrak{A}$ has a transversal note that

$$|A_1| = |A_2| = \cdots = |A_p| = n - q$$

and that, as each i is missing from $p - L(i)$ ($\leq n - q$) rows of L, it follows that each i is in at most $n - q$ of the sets $A_1, \ldots, A_p$. Hence, exactly as in the proof of the previous theorem, any r of these sets contain between them at least r different numbers. Therefore by Hall's theorem $\mathfrak{A}$ has a transversal.

To show that $\mathfrak{A}$ has a transversal **which includes the set** P we shall show that

$$\left| P \setminus \left(\bigcup_{i \in I} A_i \right) \right| \leq p - |I|$$

for each $I \subseteq \{1, \ldots, p\}$ and then use the theorem on page 31. To avoid complicated labelling we shall simply establish that inequality in the case when $I = \{1, \ldots, r\}$ (i.e. we shall only consider the union of the *first* r sets rather than of any r any sets, but the general result is very similar). In that event,

$$\left| P \setminus \left(\bigcup_{i \in I} A_i \right) \right| = \text{number of members of } P \text{ in none of } A_1, \ldots, A_r$$

$$= \text{number of members of } P \text{ in all the first } r \text{ rows of } L.$$

Now each member of P occurs precisely $p + q - n$ times in L and so the above total is zero if $r > p + q - n$. On the other hand, if $r \leq p + q - n$ then by taking $m = p + q - n$ in the lemma we get

$$\left| P \setminus \left(\bigcup_{i \in I} A_i \right) \right| = \text{number of members of } \{1, \ldots, n\} \text{ occurring exactly } p + q - n$$
$$\text{times in } L \text{ and in all the first } r \text{ rows of } L$$

$$\leq \frac{(n-q)(p-r)}{p - (p+q-n)}$$

$$= p - r$$

$$= p - |I|$$

as required. Hence, by the theorem on page 31, $\mathfrak{A}$ has a transversal which includes P.

Latin squares

We shall now use this transversal to form an extra column for L to extend it to a $p \times (q+1)$ Latin rectangle L'. If $L'(i)$ denotes the number of occurrences of i in L' then we have

$$L'(i) = L(i) + 1 = (p + q - n) + 1 \qquad \text{if } i \in P$$

and

$$L'(i) \geq L(i) > p + q - n \qquad \text{if } i \notin P.$$

Hence in each case

$$L'(i) \geq p + (q + 1) - n$$

and the process is ready to be repeated. Therefore additional columns can be added in this way until a $p \times n$ Latin rectangle is reached: the previous theorem then shows that this can be extended to an $n \times n$ Latin square as required. □

The second aspect of Latin squares which we are going to consider is the property of 'orthogonality', a concept motivated by a problem of the following type:

Example *There are 16 officers, one of each of four ranks from each of four regiments. Arrange them in a 4×4 array so that in each row and in each column there is one officer from each regiment* **and** *one officer of each rank.*

Solution Let us first ignore the ranks of the officers and simply arrange them so that in each row and column there is one from each of the regiments 1, 2, 3 and 4, as shown on the left below:

$$\begin{pmatrix} 1 & 2 & 3 & 4 \\ 2 & 1 & 4 & 3 \\ 3 & 4 & 1 & 2 \\ 4 & 3 & 2 & 1 \end{pmatrix} \qquad \begin{pmatrix} 1 & 2 & 3 & 4 \\ 3 & 4 & 1 & 2 \\ 4 & 3 & 2 & 1 \\ 2 & 1 & 4 & 3 \end{pmatrix}$$

The regiments of the officers   The ranks of the officers

So we now have four positions reserved for the officers of regiment 1, etc. We must decide which of those positions is assigned to the officers from that regiment of rank 1, 2, 3 and 4. One suitable choice of how the officers stand by rank is shown on the right above. So, for example, with this layout the first position in the second row would be taken by the officer of the third rank in the second regiment. Overall the position of the officers is shown in the following array, where the first entry in each case denotes the regiment and the second entry the rank:

$$\begin{pmatrix} 1,1 & 2,2 & 3,3 & 4,4 \\ 2,3 & 1,4 & 4,1 & 3,2 \\ 3,4 & 4,3 & 1,2 & 2,1 \\ 4,2 & 3,1 & 2,4 & 1,3 \end{pmatrix}$$

□

It is clear that the solution to that 'officers' problem consists of two Latin squares, but what is the further connection between them? When you amalgamate the two as in the final array the 16 entries consist of all the possible pairs; 1,1; 1,2; 1,3; ... ; 4,4: the two original Latin squares are called 'orthogonal'. In general two $n \times n$ Latin squares $L = (l_{ij})$ and $M = (m_{ij})$ are *orthogonal* if the n^2 pairs (l_{ij}, m_{ij}) are all different. The amalgamated array of the pairs from two orthogonal Latin squares is sometimes referred to as a *Graeco-Latin square* or as an *Euler square*. Euler's name will feature many times in this book: it occurs there because in 1782 he conjectured that it was impossible to arrange 36 officers one of each of six ranks from each of six regiments in a 6×6 array so that in each row and in each column there is one officer from each regiment and one officer of each rank. In our new terminology he was conjecturing that there is no pair of orthogonal 6×6 Latin squares. This was eventually confirmed in an exhaustive search by G. Tarry in 1900 (and various economies in the method of proof have been found since).

For which values of n do orthogonal $n \times n$ Latin squares exist? You can soon seen that the case $n = 2$ fails because the only two 2×2 Latin squares are

$$\begin{pmatrix} 1 & 2 \\ 2 & 1 \end{pmatrix} \quad \begin{pmatrix} 2 & 1 \\ 1 & 2 \end{pmatrix},$$

and we will take Tarry's word for the fact that the case $n = 6$ also fails. But what about the rest? Euler believed that it failed for all n which are even but not divisible by 4. That much wider conjecture was gradually knocked down until in 1960 it was laid to rest when R.C. Bose, S.S. Shrikhande and E.T. Parker proved in the *Canadian Journal of Mathematics* that there exist pairs of orthogonal $n \times n$ Latin squares for **all** n except $n = 2$ and $n = 6$. We now turn to the problem of finding larger collections of orthogonal Latin squares.

Example *Find four 5×5 Latin squares which are mutually orthogonal; i.e. such that any pair of them is orthogonal.*

Solution One such set is

$$L_1 = \begin{pmatrix} 1 & 2 & 3 & 4 & 5 \\ 2 & 3 & 4 & 5 & 1 \\ 3 & 4 & 5 & 1 & 2 \\ 4 & 5 & 1 & 2 & 3 \\ 5 & 1 & 2 & 3 & 4 \end{pmatrix}, \quad L_2 = \begin{pmatrix} 1 & 2 & 3 & 4 & 5 \\ 3 & 4 & 5 & 1 & 2 \\ 5 & 1 & 2 & 3 & 4 \\ 2 & 3 & 4 & 5 & 1 \\ 4 & 5 & 1 & 2 & 3 \end{pmatrix},$$

$$L_3 = \begin{pmatrix} 1 & 2 & 3 & 4 & 5 \\ 4 & 5 & 1 & 2 & 3 \\ 2 & 3 & 4 & 5 & 1 \\ 5 & 1 & 2 & 3 & 4 \\ 3 & 4 & 5 & 1 & 2 \end{pmatrix}, \quad L_4 = \begin{pmatrix} 1 & 2 & 3 & 4 & 5 \\ 5 & 1 & 2 & 3 & 4 \\ 4 & 5 & 1 & 2 & 3 \\ 3 & 4 & 5 & 1 & 2 \\ 2 & 3 & 4 & 5 & 1 \end{pmatrix}.$$

Latin squares

We note in passing that it is possible to present this collection of mutually orthogonal Latin squares as a single matrix:

$$
M = \begin{pmatrix}
i & j & L_1 & L_2 & L_3 & L_4 \\
1 & 1 & 1 & 1 & 1 & 1 \\
1 & 2 & 2 & 2 & 2 & 2 \\
1 & 3 & 3 & 3 & 3 & 3 \\
1 & 4 & 4 & 4 & 4 & 4 \\
1 & 5 & 5 & 5 & 5 & 5 \\
2 & 1 & 2 & 3 & 4 & 5 \\
2 & 2 & 3 & 4 & 5 & 1 \\
2 & 3 & 4 & 5 & 1 & 2 \\
2 & 4 & 5 & 1 & 2 & 3 \\
2 & 5 & 1 & 2 & 3 & 4 \\
3 & 1 & 3 & 5 & 2 & 4 \\
3 & 2 & 4 & 1 & 3 & 5 \\
3 & 3 & 5 & 2 & 4 & 1 \\
3 & 4 & 1 & 3 & 5 & 2 \\
3 & 5 & 2 & 4 & 1 & 3 \\
4 & 1 & 4 & 2 & 5 & 3 \\
4 & 2 & 5 & 3 & 1 & 4 \\
4 & 3 & 1 & 4 & 2 & 5 \\
4 & 4 & 2 & 5 & 3 & 1 \\
4 & 5 & 3 & 1 & 4 & 2 \\
5 & 1 & 5 & 4 & 3 & 2 \\
5 & 2 & 1 & 5 & 4 & 3 \\
5 & 3 & 2 & 1 & 5 & 4 \\
5 & 4 & 3 & 2 & 1 & 5 \\
5 & 5 & 4 & 3 & 2 & 1
\end{pmatrix}
$$

← For example this row of M has the information about the (2,4)th entry in each of the Latin squares: in L_1 it is 5, in L_2 it is 1, in L_3 it is 2 and in L_4 it is 3.

The order of the rows of M is irrelevant: we have merely presented them in a fairly natural order. Now note that M has a special property: nowhere in it will you find a rectangle of entries of the form

$$
\begin{matrix}
x & \cdots & y \\
\vdots & & \vdots \\
x & \cdots & y
\end{matrix}
$$

For example, such a rectangle of entries in columns 1 and 4 would mean that in row x the Latin square L_2 had two entries of y: and such a rectangle in columns 3 and 5 would contradict the orthogonality of L_1 and L_3. You can quickly think through the other cases for yourselves.

In general any r mutually orthogonal $n \times n$ Latin squares with entries in $\{1, \ldots, n\}$ can be given as a single $n^2 \times (r+2)$ matrix M in this way, where the entries of M are from $\{1, \ldots, n\}$ and where M has no rectangle of entries of the above form. Indeed the converse is also true: given any $n^2 \times (r+2)$ matrix M with entries in $\{1, \ldots, n\}$ and with no rectangle of entries of the above form, then labelling the columns of M as $i, j, L_1, \ldots, L_r$ means that each row of M can be used to define the (i, j)th entries of some $n \times n$ matrices $L_1, \ldots, L_r$. The special 'non-rectangle' property of M ensures that all values of (i, j) are covered, that the resulting matrices are Latin squares, and that they are all mutually orthogonal. Again it is left as an exercise to confirm these facts. □

A natural question is, for any given $n > 1$, what is the largest possible number of mutually orthogonal $n \times n$ Latin squares? The next theorem shows that it is never possible to find n such squares.

Theorem *For any integer $n > 1$ there exist at most $n - 1$ mutually orthogonal $n \times n$ Latin squares.*

Proof Let $L_1, \ldots, L_q$ be a collection of mutually orthogonal $n \times n$ Latin squares: we aim to show that $q \leq n - 1$. Note that in the previous example the first row of each of the Latin square was (1 2 3 4 5) and we show now that it is possible to doctor $L_1, \ldots, L_q$ to give a collection of mutually orthogonal $n \times n$ Latin squares $L'_1, \ldots, L'_q$ in which the first row of each is (1 2 ... n). In L_1 assume that the first row is $(a_1 \ a_2 \ \ldots \ a_n)$: throughout L_1 replace a_1 by 1 wherever it occurs, replace a_2 by 2 wherever it occurs, ... and replace a_n by n wherever it occurs, and call the resulting array L'_1. It is straightforward to check that L'_1 is still a Latin square and that it is still orthogonal to each of $L_2, \ldots, L_q$. Repeating this procedure independently for each of the other Latin squares gives the mutually orthogonal Latin squares $L'_1, \ldots, L'_q$ with first rows (1 2 ... n) as claimed:

$$L'_1 = \begin{pmatrix} 1 & 2 & 3 & \ldots & n \\ ? & & & & \\ & & & & \end{pmatrix}, L'_2 = \begin{pmatrix} 1 & 2 & 3 & \ldots & n \\ ? & & & & \\ & & & & \end{pmatrix}, \ldots, L'_q = \begin{pmatrix} 1 & 2 & 3 & \ldots & n \\ ? & & & & \\ & & & & \end{pmatrix}.$$

What are the entries marked '?' in these q Latin squares? None of them is 1 because there is already a 1 in the first column of each square. Furthermore, no two of the ?s are the same because if we had, for example,

$$L'_j = \begin{pmatrix} 1 & 2 & 3 & \ldots & i & \ldots & n \\ & & & & i & & \end{pmatrix}, \quad L'_k = \begin{pmatrix} 1 & 2 & 3 & \ldots & i & \ldots & n \\ & & & & i & & \end{pmatrix},$$

then across these two squares the pair i, i would appear twice, contradicting the orthogonality of L'_j and L'_k. Hence the q entries marked '?' are *different* members of $\{2, 3, \ldots, n\}$ and it follows that $q \leqslant n - 1$ is required. □

So the four mutually orthogonal 5×5 Latin squares displayed in the above example is the largest such set. The neat pattern of those four leads us to suspect that for some n there will be a general construction of $n - 1$ mutually orthogonal Latin squares, and this turns out to be the case when n is prime (and when n is a power of a prime, but that requires some more advanced algebra).

Theorem *If n is a prime or a power of a prime then there exist $n - 1$ mutually orthogonal $n \times n$ Latin squares.*

Proof Until now our $n \times n$ Latin squares have had entries from $\{1, 2, \ldots, n\}$ but in this proof it is more convenient to choose the entries from $\{0, 1, \ldots, n - 1\}$. In modular arithmetic the set of numbers $\{0, 1, 2, \ldots, n - 1\}$ can be added and multiplied in a fairly sensible way 'mod n' to give answers back in the same set. So, for example, the addition and multiplication tables 'mod 5' are as shown:

+	0	1	2	3	4
0	0	1	2	3	4
1	1	2	3	4	0
2	2	3	4	0	1
3	3	4	0	1	2
4	4	0	1	2	3

·	0	1	2	3	4
0	0	0	0	0	0
1	0	1	2	3	4
2	0	2	4	1	3
3	0	3	1	4	2
4	0	4	3	2	1

Addition mod 5   Multiplication mod 5

Essentially the arithmetic operations are the normal ones but then the answers are reduced 'mod n' by taking away all possible multiples of n. (If you have studied sufficient abstract algebra you will know that in the case when n is prime the above process defines a 'field'.)

Now let n be prime and let $+$ and $\cdot$ denote addition and multiplication 'mod n'. Define a collection $L_1, \ldots, L_{n-1}$ of $n \times n$ matrices by the rule that the (i, j)th entry of L_k is $k \cdot (i - 1) + (j - 1)$. (If you do this in the case $n = 5$ you will get four 5×5 Latin squares very closely related to the collection displayed in the earlier example.) Clearly this process defines a collection of $n \times n$ matrices with entries chosen from $\{0, 1, \ldots, n - 1\}$: we now show that this collection consists of $n - 1$ mutually orthogonal $n \times n$ Latin squares.

(i) L_k has no repeated entry in column j.

If the (i, j)th entry equals the (i', j)th entry in L_k, where $i > i'$, then

$$k \cdot (i - 1) + (j - 1) = k \cdot (i' - 1) + (j - 1)$$

and (as $+$ and $\cdot$ behave very naturally) we can deduce that

$$k \cdot (i - 1) = k \cdot (i' - 1).$$

This means that $k(i - 1)$ and $k(i' - 1)$ give the same remainder when divided by n and that their difference, $k(i - i')$, is divisible by n. Since n is prime we can deduce that either k or $i - i'$ is divisible by n. But as $1 \leq k \leq n - 1$ and $1 \leq i - i' \leq n - 1$ this is clearly impossible. Hence no L_k has a repeated entry in any of its columns.

(ii) L_k has no repeated entry in row i.

This is very similar to (i), but slightly easier, and is left as an exercise.

(iii) Properties (i) and (ii) confirm that each L_k is a Latin square. We now show that L_k and $L_{k'}$ are orthogonal.

If $k \neq k'$ and L_k and $L_{k'}$ are not orthogonal then across these two squares some pair will occur twice, in the (i, j)th and (i', j')th places, say.

$$L_k = \begin{pmatrix} & x & \\ & & x \end{pmatrix} \begin{matrix} \leftarrow i \\ \leftarrow i' \end{matrix} \qquad L_{k'} = \begin{pmatrix} & y & \\ & & y \end{pmatrix} \begin{matrix} \leftarrow i \\ \leftarrow i' \end{matrix}$$

$$\begin{matrix} \uparrow & \uparrow \\ j & j' \end{matrix} \qquad\qquad \begin{matrix} \uparrow & \uparrow \\ j & j' \end{matrix}$$

But then the (i, j)th entry of L_k will equal the (i', j')th entry of L_k and the (i, j)th entry of $L_{k'}$ will equal the (i', j')th entry of $L_{k'}$; i.e.

$$k \cdot (i - 1) + (j - 1) = k \cdot (i' - 1) + (j' - 1)$$

and

$$k' \cdot (i - 1) + (j - 1) = k' \cdot (i' - 1) + (j' - 1).$$

Subtracting these two equations shows us that

$$(k - k')(i - i')$$

is divisible by n, but an argument similar to that in (i) using the primeness of n shows that this is impossible. Hence L_k and $L_{k'}$ are indeed orthogonal.

We have therefore seen how to construct $n - 1$ mutually orthogonal $n \times n$ Latin squares in the case when n is prime. We now give a brief outline of how to extend this to the case when n is a power of a prime: it needs rather more abstract algebra and some readers may prefer to omit the rest of this proof.

In our construction above the key fact about $\{0, 1, \ldots, n-1\}$ under $+$ and $\cdot$ is that it forms a field (essentially the operations behave in a sensible arithmetic fashion and, in particular, if the product of two numbers is zero then one of the numbers must itself be zero). In the case when n is a power of a prime this set-up fails to be a field (for example if $n = p^r$ where $r > 1$ then $p \cdot p^{r-1} = 0$). However in this case it is still possible to define other operations of addition and multiplication on $\{0, 1, \ldots, n-1\}$ which make it into a field: this field is known as the *Galois field* $GF(n)$. The details of the operations (defined *via* polynomials) need not concern us here but once we know that such a field exists the first part of this proof can easily be generalised to the case when n is a power of a prime. □

In that proof we essentially saw that if there exists a field of n elements then there exist $n - 1$ mutually orthogonal $n \times n$ Latin squares (but not conversely) and it is known that such a field exists precisely when n is a prime or a power of a prime. But it still remains an unsolved conjecture that there exists a 'full set' of $n - 1$ mutually orthogonal $n \times n$ Latin squares **if and only if** n is a prime or a power of a prime. We shall return to this conjecture shortly.

Our final topic in this chapter seems, at first sight, to be unrelated to Latin squares. Early this century Oswald Veblen and others considered an abstract form of 'geometry' of which the following is a typical example:

Example *Note that in this figure there are seven 'points', seven 'lines' (including the circular one) and that they have the properties:*
(i) *any two points lie on precisely one line;*
(i)′ *any two lines meet in precisely one point;*
(ii) *each line contains three points;*
(ii)′ *each point lies on three lines.*
So this example consists of a set of 'points' $\{1, 2, 3, 4, 5, 6, 7\}$ and a collection of 'lines' $\{1, 2, 3\}$, $\{1, 4, 5\}$, $\{1, 6, 7\}$, $\{2, 4, 7\}$, $\{2, 5, 6\}$, $\{3, 4, 6\}$ and $\{3, 5, 7\}$.

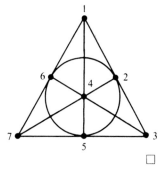

□

In general a *finite projective plane (of order n)* consists of a finite set (whose elements are called the *points*) and a collection of subsets of that set (each subset being called a *line*) satisfying
(i) any two points lie on precisely one line;
(i)′ any two lines meet in precisely one point;
(ii) each line contains $n + 1$ points;
(ii)′ each point lies on $n + 1$ lines.
So the previous example was of a finite projective plane of order 2. In the exercises we shall see that a finite projective plane of order n must have precisely $n^2 + n + 1$

points. In fact these geometries have a more minimal set of axioms and some fascinating properties of 'duality' obtained by interchanging the roles of the points and the lines. We shall see some of these properties in the more modern context of a block design in chapter 15, and more details can be found, for example, in H. J. Ryser's *Combinatorial mathematics* listed in the bibliography.

But what is the connection between finite projective planes and Latin squares? It is known that finite projective planes of order n exist if n is prime or is a power of a prime and it has been conjectured that these are the only ones which exist: this is very reminiscent of the conjecture concerning the existence of $n - 1$ mutually orthogonal $n \times n$ Latin squares, and we now see that these two existence problems are equivalent.

Theorem *There exists a finite projective plane of order n if and only if there exist $n - 1$ mutually orthogonal $n \times n$ Latin squares.*

Sketch proof We shall merely illustrate the connection between the two problems by constructing a finite projective plane of order 3 from two given orthogonal 3×3 Latin squares and, conversely, by constructing two orthogonal 3×3 Latin squares from a given finite projective plane of order 3. Of course this does not constitute a proof of the general result but the techniques do generalise easily, and we include some comments about the general case.

($\Leftarrow$) Start with two orthogonal 3×3 Latin squares:

$$L_1 = \begin{pmatrix} 3 & 1 & 2 \\ 2 & 3 & 1 \\ 1 & 2 & 3 \end{pmatrix} \qquad L_2 = \begin{pmatrix} 2 & 1 & 3 \\ 1 & 3 & 2 \\ 3 & 2 & 1 \end{pmatrix}.$$

Then write them as one combined matrix as we did earlier:

$$M = \begin{pmatrix} 1 & 1 & 3 & 2 \\ 1 & 2 & 1 & 1 \\ 1 & 3 & 2 & 3 \\ 2 & 1 & 2 & 1 \\ 2 & 2 & 3 & 3 \\ 2 & 3 & 1 & 2 \\ 3 & 1 & 1 & 3 \\ 3 & 2 & 2 & 2 \\ 3 & 3 & 3 & 1 \end{pmatrix} \quad \leftarrow \text{e.g. the } (1,3)\text{rd entry of } L_1 \text{ is 2 and of } L_2 \text{ is 3.}$$

Now introduce a set of 13 points $\{c_1, c_2, c_3, c_4, r_1, r_2, r_3, r_4, r_5, r_6, r_7, r_8, r_9\}$ (which can be thought of as referring to the columns 1–4 and the rows 1–9). Then consider

Latin squares

'lines' formed by the following subsets of four of those points:

$$\{c_1, c_2, c_3, c_4\}$$

and any of the form

$$\{c_j, r_s, r_t, r_u\}$$

where the three entries in M in the jth column and in rows s, t and u are the same. For example one of these sets will be $\{c_2, r_3, r_6, r_9\}$ because the entries in rows 3, 6 and 9 of column 2 are all the same (namely 3). Overall this gives the following 13 'lines':

$$\{c_1, c_2, c_3, c_4\} \quad \{c_1, r_1, r_2, r_3\} \quad \{c_1, r_4, r_5, r_6\} \quad \{c_1, r_7, r_8, r_9\}$$
$$\{c_2, r_1, r_4, r_7\} \quad \{c_2, r_2, r_5, r_8\} \quad \{c_2, r_3, r_6, r_9\} \quad \{c_3, r_2, r_6, r_7\} \quad \{c_3, r_3, r_4, r_8\}$$
$$\{c_3, r_1, r_5, r_9\} \quad \{c_4, r_2, r_4, r_9\} \quad \{c_4, r_1, r_6, r_8\} \quad \{c_4, r_3, r_5, r_7\}.$$

It is now straightforward to check that these 13 points and sets satisfy the axioms of a finite projective plane in the case $n = 3$. In general the $n - 1$ mutually orthogonal $n \times n$ Latin squares will give an $n^2 \times (n + 1)$ matrix M with entries in $\{1, \ldots, n\}$ and with no rectangle of entries of the form

$$\begin{matrix} x & \ldots & y \\ \vdots & & \vdots \\ x & \ldots & y \end{matrix}$$

The above construction will then give $n^2 + n + 1$ points $\{c_1, \ldots, c_{n+1}, r_1, \ldots, r_{n^2}\}$ and $n^2 + n + 1$ lines each containing $n + 1$ points and such that each pair of points lies in just one line. In general the non-rectangle property of M will ensure that these points and lines satisfy the axioms of a finite projective plane. For example how many points will be in both the lines

$$\{c_j, r_i, \ldots\} \quad \text{and} \quad \{c_{j'}, r_{i'}, \ldots\}?$$

If $j = j'$ then the c_j is clearly the only point in common. And if $j \neq j'$ then the first line will have resulted from all the rows containing a '1' say in the jth column, and the second line will have resulted from all the rows containing a '2' say in the j'th column. The non-rectangle property of M ensures that the pair $(1, 2)$ occurs precisely once across the columns j and j' (in the ith row, say, as shown) and hence that the two given lines intersect in the one point r_i.

($\Rightarrow$) Conversely assume that we are given a finite projective plane of order 3. It will consist of 13 points and 13 lines, with each line consisting of 4 points. Label the points of one of the lines as c_1, c_2, c_3, c_4 and label the remaining points as $r_1, r_2, r_3, r_4, r_5, r_6, r_7, r_8$ and r_9. Then, for example, the lines might be:

$$\begin{pmatrix} & & 2 & \\ & 1 & & \\ & & 2 & \\ & 1 & 2 & \leftarrow i \\ & 1 & & \\ & & 2 & \\ & 1 & & \\ & \uparrow & \uparrow & \\ & j & j' & \end{pmatrix}$$

$$\{c_1, c_2, c_3, c_4\} \quad \{c_1, r_1, r_2, r_3\} \quad \{c_1, r_4, r_5, r_6\} \quad \{c_1, r_7, r_8, r_9\}$$
$$\{c_2, r_1, r_4, r_7\} \quad \{c_2, r_2, r_5, r_8\} \quad \{c_2, r_3, r_6, r_9\} \quad \{c_3, r_2, r_6, r_7\} \quad \{c_3, r_3, r_4, r_8\}$$
$$\{c_3, r_1, r_5, r_9\} \quad \{c_4, r_2, r_4, r_9\} \quad \{c_4, r_1, r_6, r_8\} \quad \{c_4, r_3, r_5, r_7\}.$$

The fact that any two of the lines meet in a single point means that, apart from the line $\{c_1, c_2, c_3, c_4\}$, the remaining 12 lines are bound to fall into four groups of three as follows:

$$\begin{array}{llll}
\text{containing } c_1: & \{c_1, r_1, r_2, r_3\} & \{c_1, r_4, r_5, r_6\} & \{c_1, r_7, r_8, r_9\} \\
\text{containing } c_2: & \{c_2, r_1, r_4, r_7\} & \{c_2, r_2, r_5, r_8\} & \{c_2, r_3, r_6, r_9\} \\
\text{containing } c_3: & \{c_3, r_2, r_6, r_7\} & \{c_3, r_3, r_4, r_8\} & \{c_3, r_1, r_5, r_9\} \\
\text{containing } c_4: & \{c_4, r_2, r_4, r_9\} & \{c_4, r_1, r_6, r_8\} & \{c_4, r_3, r_5, r_7\} \\
& 1 & 2 & 3
\end{array}$$

Call the first set in each row '1', the second set '2' and the third set '3', as shown. Then define a 9 × 4 matrix M by the rule that the (i, j)th entry is k if the pair $\{c_j, r_i\}$ lies in a set labelled k. In our example this gives rise to the matrix

$$M = \begin{pmatrix} 1 & 1 & 3 & 2 \\ 1 & 2 & 1 & 1 \\ 1 & 3 & 2 & 3 \\ 2 & 1 & 2 & 1 \\ 2 & 2 & 3 & 3 \\ 2 & 3 & 1 & 2 \\ 3 & 1 & 1 & 3 \\ 3 & 2 & 2 & 2 \\ 3 & 3 & 3 & 1 \end{pmatrix} \quad \text{e.g. } \{c_4, r_5\} \text{ lies in the set number 3}$$

We can then use this matrix to read off, in the usual way, the two orthogonal 3 × 3 Latin squares

$$L_1 = \begin{pmatrix} 3 & 1 & 2 \\ 2 & 3 & 1 \\ 1 & 2 & 3 \end{pmatrix} \quad L_2 = \begin{pmatrix} 2 & 1 & 3 \\ 1 & 3 & 2 \\ 3 & 2 & 1 \end{pmatrix}.$$

This process will work in general: the finite projective plane will consist of $n^2 + n + 1$ points and lines and will give rise to an $n^2 \times (n + 1)$ matrix M with entries in $\{1, \ldots, n\}$. The finite projective plane axioms will ensure that the matrix M has the usual non-rectangle property because entries of the form

$$i \to x \ \ldots \ y$$

$$i' \to x \ \ldots \ y$$

would mean that $\{r_i, r_{i'}\}$ lies in two of the lines. Hence M will give rise to $n - 1$ mutually orthogonal $n \times n$ Latin squares as required. □

So finite projective planes of orders $n = 2, 3, 4, 5, 7, 8, 9$, and 11 all exist because these are all primes or powers of primes and so are covered by our existence theorem for $n - 1$ mutually orthogonal $n \times n$ Latin squares. There is no finite projective plane of order 6 because, as we observed earlier, there exists no pair of orthogonal 6×6 Latin squares. (We shall also obtain our own proof of the non-existence of a finite projective plane of order 6 in chapter 15.) The next case not covered by our theorems is $n = 10$ and, remarkably, it was only in 1989 that C. Lam, L. Thiel and S. Swierz were able to announce that an exhaustive computer search (taking over 3000 hours) had shown that no finite projective plane of order 10 exists.

Exercises

1. For which values of i and j can the following Latin rectangles with entries in $\{1, 2, 3, 4, 5, 6\}$ be extended to 6×6 Latin squares? Carry out one such extension in each possible case.

$$\begin{pmatrix} 1 & 2 & 3 & 4 \\ 5 & 6 & 1 & 2 \\ 3 & 4 & 5 & 1 \\ 4 & 1 & 2 & i \end{pmatrix} \qquad \begin{pmatrix} 1 & 2 & 3 & 4 \\ 5 & 1 & 2 & 6 \\ 3 & 4 & 5 & 1 \\ 4 & 3 & 1 & j \end{pmatrix}. \qquad \text{[A]}$$

2. Show that the number of $n \times n$ Latin squares with entries in $\{1, \ldots, n\}$ is at least $1! \times 2! \times \cdots \times n!$. [H]

3. Let L be a $p \times q$ Latin rectangle with entries in $\{1, \ldots, n\}$ in which each member of $\{1, \ldots, n\}$ occurs in L the same number of times. Show that L can be extended to an $n \times n$ Latin square.

4. Let p, q and n be fixed integers with $1 \leq p, q < n$. Then any $p \times q$ Latin rectangle with entries in $\{1, \ldots, n\}$ can be regarded as a $p \times q$ Latin rectangle with entries in $\{1, \ldots, N\}$ for any $N \geq n$. Show that any $p \times q$ Latin rectangle can be extended to a Latin square of some size $N \times N$. What (in terms of p, q and n) is the smallest value of N which will work for all $p \times q$ Latin rectangles with entries in $\{1, \ldots, n\}$? [H,A]

5. Show that if L and L^T (= the transpose of L) are orthogonal $n \times n$ Latin squares then the leading diagonal of L has no repeated entry. Deduce that no such L exists in the case $n = 3$ and find such an L in the case $n = 4$. [A]

6. (i) Use the modular arithmetic construction from the proof of the theorem on page 63 to find four mutually orthogonal 5×5 Latin squares. (If you are feeling keen you could then construct a finite projective plane from them, using the construction in the last proof.) [A]

(ii) Show that for general n that modular arithmetic construction will give $n \times n$ matrices $L_1, \ldots, L_{n-1}$ but that L_k will be a Latin square if and only if k and n are relatively prime. Show also that L_k and $L_{k'}$ will be orthogonal Latin squares if and only if n is relatively prime to each of k, k' and $k - k'$. [H]

(iii) Let the operations of $+$ and $\cdot$ be defined on $\{0, 1, 2, 3\}$ by

+	0	1	2	3
0	0	1	2	3
1	1	0	3	2
2	2	3	0	1
3	3	2	1	0

$\cdot$	0	1	2	3
0	0	0	0	0
1	0	1	2	3
2	0	2	3	1
3	0	3	1	2

(These make $\{0, 1, 2, 3\}$ into a field.) Use these operations to construct three mutually orthogonal 4×4 Latin squares. [A]

7. Show that a finite projective plane of order n contains $n^2 + n + 1$ points and $n^2 + n + 1$ lines. [H]

6

The first theorem of graph theory

In the eighteenth century the inhabitants of Königsberg in Eastern Prussia (at time of writing – though not necessarily at time of reading! – this is Kaliningrad in the Russian Federation) apparently puzzled over the now-famous problem of whether they could go on a walking tour of their city crossing each of the seven bridges once and only once.

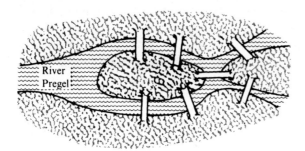

Königsberg

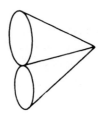

Schematic representation

Representing each bridge by a line and collapsing each land mass to a single point we are able to see that this is equivalent to drawing the right-hand figure in one continuous piece without covering any stretch of line twice. As we observed by a parity argument in chapter 4 it is impossible to draw a figure in such a way if it has more than two points from which an odd number of lines emanate. Since this particular figure has four 'odd' junctions it is impossible to find the required route. It was in 1736, in a paper by Leonhard Euler, that such a mathematical argument was first formulated: a translation of his paper can be found in *Graph theory 1736–1936*, which has already been recommended earlier, and in fact the front cover of the paperback edition consists of a seventeenth century artist's impression of Königsberg. The former date in the title of that book reflects the fact that Euler's paper is generally acknowledged as the first treatise on what is now called graph theory. In this chapter we shall look at Euler's result and at other related theorems concerning paths in graphs which, in some sense, complete a tour of the graph.

The first task, motivated by the Königsberg bridge problem, is to characterise those graphs which have a path using all their edges.

Example *Can the left-hand figure be drawn without taking your pencil off the paper and without covering any stretch of line twice? Equivalently, does the graph on the right have a path which uses all its edges?*

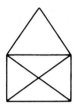
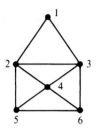

Solution It is easy to draw the left-hand figure as required or, equivalently, to note for example that the path 5, 2, 1, 3, 2, 4, 5, 6, 4, 3, 6 uses all the edges in the graph. ☐

Note that our observation about odd junctions means that graphs with more than two vertices of odd degree cannot possess a path using all their edges. Also, for such a path to exist we certainly need the graph to be connected. But we cannot deduce from those observations that if a connected graph has two or fewer vertices of odd degree (like the one in the previous example) then it necessarily has a path using all its edges. Although Euler implied in his 1736 paper that this converse was also true it was not actually proved until over a century later: even so this characterisation of such graphs in terms of the degrees of the vertices is generally known as *Euler's theorem* or as 'the first theorem of graph theory'. We start with the case where the path is required to finish where it started: in general such a path is called a *closed path*, a closed path which uses all the edges of the graph is called an *Eulerian* path, and a graph which possesses an Eulerian path is itself called *Eulerian*.

Theorem (Euler/Hierholzer) *Let G be a connected graph. Then the following three properties are equivalent:*
(i) *each vertex of G has even degree;*
(ii) *G has some cycles which between them use each edge of G once and only once;*
(iii) *G is Eulerian.*

Proof (i) ⇔ (ii).   We have already seen that (i) implies (ii) in exercise 3 on page 23 (and in fact it does not rely on G being connected). For if (i) holds then G has no vertex of degree 1. So if $E \neq \emptyset$ then G has a component which is not a tree and so it possesses a cycle. Removing the edges of that cycle leaves a graph in which each degree is still even and the process can be repeated until sufficient cycles have been removed to use up all the edges of G. This argument can easily be made more formal by the use of mathematical induction. The converse that (ii) implies (i) is immediate, for if a vertex lies on r of the given cycles then it clearly has degree $2r$.

(ii) ⇒ (iii). It is intuitively clear that if G is connected and consists of a disjoint union of cycles as described in (ii) then those cycles can be strung together in some way to form an Eulerian path (i.e. a closed path using all the edges of G). To make this argument more formal, induction is once again needed. Assume that (ii) holds and that G has r cycles which between them use all the edges of G once and only once. We shall deduce by induction on r that G has an Eulerian path. The case $r = 0$ is trivial since then the connected graph G will consist of a single vertex, v say, and the trivial closed path consisting of v alone will use all the edges of G. So assume that $r > 0$ and that the result is known for graphs which are disjoint unions of fewer than r cycles. As G is connected, clearly we can label the edge-sets of the r cycles as $C_1, \ldots, C_r$ such that for each i the graph consisting of edge-set $C_1 \cup \cdots \cup C_i$ and vertex-set consisting of the endpoints of those edges is connected.

e.g.

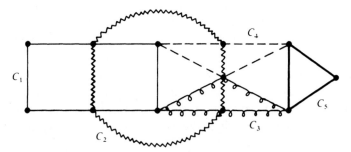

Then in particular the graph with edge-set $C_1 \cup \cdots \cup C_{r-1}$ and vertex-set consisting of the endpoints of those edges is connected and has each degree even: therefore by the induction hypothesis that graph has an Eulerian path. Since G is connected, at some stage that path will use a vertex which is also the endpoint of an edge in C_r and so a round detour using the edges of C_r can be made. This will give an Eulerian path in G, as required.

(iii) ⇒ (i). The proof that (iii) implies (i) is essentially our earlier observation about 'odd' junctions. Formally assume (iii) and let $G = (V, E)$ have an Eulerian path

$$v_1, v_2, v_3, \ldots, v_n, v_1$$

Then E consists of the n edges

$$E = \{v_1v_2, v_2v_3, \ldots, v_nv_1\}.$$

It follows easily that if a vertex v of G occurs r times in the list $v_1, v_2, \ldots, v_n$ then v has degree $2r$ in G. Hence each vertex of G has even degree and (i) follows. □

A non-closed path which uses all the edges of a graph is called a *semi-Eulerian* path (and a graph which possesses one is called *semi-Eulerian*). We can now deduce the corresponding result about semi-Eulerian graphs.

Corollary *Let G be a connected graph. Then G is semi-Eulerian if and only if G has precisely two vertices of odd degree.*

Proof ($\Rightarrow$) Assume that $G = (V, E)$ has a semi-Eulerian path $v_1, v_2, \ldots, v_n$ (i.e. $v_1 \neq v_n$ and it uses all the edges of G). Then let v^* be a completely new vertex and consider the graph $G^* = (V \cup \{v^*\}, E \cup \{v_n v^*, v^* v_1\})$ as illustrated:

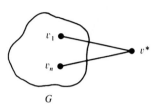

G

Clearly the path $v_1, v_2, \ldots, v_n, v^*, v_1$ is an Eulerian path in G^*. It follows from the theorem that each vertex of G^* has even degree. Removing the two edges $v_n v^*$ and $v^* v_1$ reduces the degrees of v_n and v_1 by one and leaves the remaining degrees in G unaffected. It follows that G has precisely two vertices of odd degree, namely v_n and v_1.

($\Leftarrow$) Conversely assume that the connected graph $G = (V, E)$ has n vertices precisely two of which have odd degree: call them v_1 and v_n. Then construct G^* from G exactly as above. It is clear that G^* is connected and has each vertex of even degree. Hence by the theorem G^* has an Eulerian path. On such a 'round walk' we can start anywhere and go in either direction, so G^* has an Eulerian path of the form:

$$v^*, v_1, v_2, v_3, \ldots, v_n, v^*.$$

It is then clear that the path

$$v_1, v_2, v_3, \ldots, v_n$$

is a semi-Eulerian path in G. □

Examples

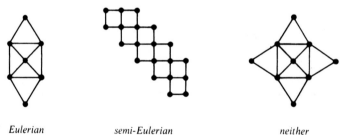

Eulerian   semi-Eulerian   neither

Given a particular connected graph in which each vertex has even degree it is possible to use the construction from the proof of the theorem to actually find an Eulerian path. For it is easy to find and remove the edges of a cycle from the graph and to keep repeating the process until all the edges have gone. Then the cycles can be put back together to form the required Eulerian path.

Example

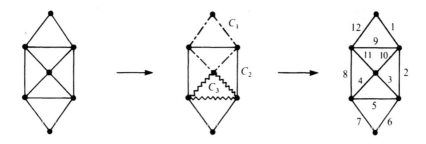

Go around C_1 and detour around each other cycle when first met. □

However a much more systematic way of constructing an Eulerian path is given by the following algorithm. It requires the idea of an edge which 'disconnects' a graph: naturally enough, an edge e in the connected graph $G = (V, E)$ *disconnects* G if the graph $(V, E\backslash\{e\})$ is disconnected (and such an edge is sometimes called an *isthmus* or *bridge*). Basically the algorithm says proceed along any path you like, rubbing out edges as you use them and also rubbing out any vertices of degree zero left behind you: the only restriction is that you must only use a disconnecting edge if there is no other choice.

Theorem (Fleury's algorithm) *Given a connected graph G in which each vertex has even degree construct a path by the following process.*
  I. *Write down any vertex v_1 and let the graph G_1 be G itself.*
  II. *Suppose that vertices $v_1, \ldots, v_i$ and graph G_i have already been chosen. Then*
    (i) *if G_i has no edges of the form $v_i v$ then stop;*
    (ii) *if G_i has an edge $v_i v$ which does not disconnect G_i then let v_{i+1} be any such v, let G_{i+1} be G_i with the edge $v_i v_{i+1}$ deleted and with the vertex v_i deleted if it then has zero degree; then repeat stage II;*
    (iii) *if each edge of G_i of the form $v_i v$ disconnects G_i then let v_{i+1} be any such v, let G_{i+1} be G_i with the edge $v_i v_{i+1}$ deleted and with the vertex v_i deleted if it then has zero degree; then repeat stage II.*
It follows that the resulting sequence of vertices will be an Eulerian path in G.

Proof Let G be as stated and apply the algorithm to it. At each stage G_i consists of G with edges $v_1 v_2, \ldots, v_{i-1} v_i$ deleted and with any of the vertices $v_1, \ldots, v_{i-1}$ deleted if they then have zero degree. So it is easy to see that in each G_i each vertex except possibly for v_1 and v_i has even degree, and those two have odd degree if $v_1 \neq v_i$ and even degree if $v_1 = v_i$. Furthermore, with the possible exception of v_i, each vertex of G_i has positive degree.

Suppose that the algorithm stops at G_i: then it has certainly produced a path $v_1, \ldots, v_i$ in G and it must have stopped because there is no edge of the form $v_i v$ in

G_i. So the vertex v_i has zero degree in G_i and, as this degree is even, the above comments show that $v_i = v_1$ and that the path produced is closed. If it has used all the edges of G then it is the required Eulerian path. On the other hand, if some edge of G has not been used in the path, then G_i will have a non-empty edge-set and it must be of the following form:

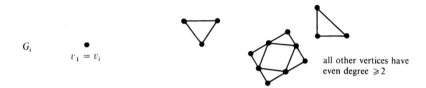

Since G was connected at least one of the vertices $v_2, \ldots, v_{i-1}$ must remain in G_i and so let v_j be the last such vertex. Then the earlier graph G_j must be of the following form:

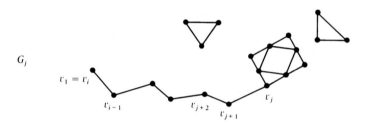

But then how has the algorithm proceeded from G_j to G_{j+1}? It has used the edge $v_j v_{j+1}$ which disconnects G_j when there was a non-disconnecting option available. This contradicts the fact that the path was generated by the algorithm. Hence the algorithm is bound to produce a closed path which uses *all* the edges of G, and the theorem is proved. □

Example *For the graph illustrated one of the possible Eulerian paths generated by Fleury's algorithm is shown, with the edges numbered in the order in which they are used: follow it for yourself (rubber in hand!) and note the two occasions ($3 \to 4$ and $8 \to 9$) when you have to avoid a disconnecting edge.*

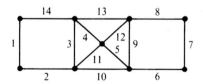

□

We now turn to a different problem where, instead of looking for some round route using all the *edges* of a graph, we are going to be looking for a special sort of round route using all the *vertices*.

The first theorem of graph theory

Example *Label the vertices of the following graph* 1, 2, ..., 20 *so that* 1 *is joined* 2, 2 *is joined to* 3, ..., 19 *is joined to* 20 *and* 20 *is joined to* 1.

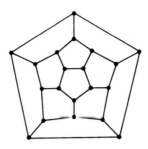

I leave you to do that easy exercise. It was the basis of *The icosian game* designed and sold by the famous algebraist Sir William Rowan Hamilton in 1856. In graph-theoretic terms you are looking for a cycle which uses all the vertices: such cycles are now known as *Hamiltonian cycles*, and a graph which possesses a Hamiltonian cycle is itself called *Hamiltonian*. We met this idea in exercise 11 on page 48, where we also met the following example:

Example *Find a route around a chess-board for a knight so that he visits each square exactly once before finishing on the same square that he started on.*

Solution If the squares of the board are regarded as the vertices of a graph, with two of them joined by an edge if it is possible for a knight to make one move from one to the other, then finding a knight's route is equivalent to finding a Hamiltonian cycle. One of them many such cycles is illustrated below:

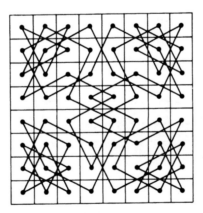

In the case of Eulerian paths we were able to give a very neat characterisation of those graphs which possessed one, but there is no known simple characterisation of those graphs which possess a Hamiltonian cycle. There are various theorems which give sufficient conditions for a Hamiltonian cycle to exist: these conditions usually involve requiring that, in some sense, the graph has plenty of edges, and we now give one such result. (It is easy to see, by considering a graph consisting of a single cycle of more than four edges, that the converse of the result is false.)

Theorem *Let $G = (V, E)$ be a graph with $|V| \geq 3$ and such that $\delta u + \delta v \geq |V|$ for each pair of distinct unjoined vertices u and v. Then G is Hamiltonian.*

Proof Assume that the theorem fails for some graph on n (≥ 3) vertices. So there exists a graph on n vertices with the following two properties:

$\delta u + \delta v \geq n$ for each unjoined pair of vertices	there is no Hamiltonian cycle

Add as many edges as possible to this graph without producing a Hamiltonian cycle. Then both of the above properties are maintained but we end up with a largest such graph G on n vertices. Clearly G is not the complete graph on n (≥ 3) vertices, for that would have a Hamiltonian cycle. So G has a pair of vertices, u and v say, which are not joined by an edge. By the choice of G if any further edges are added a Hamiltonian cycle will be created. Hence the addition of the edge uv would create a Hamiltonian cycle and it follows that G has a path from u to v which uses all n vertices of G just once:

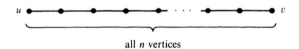

all n vertices

Put a circle around the δu vertices to which u is joined and put a square around each vertex one place to the left of a circle:

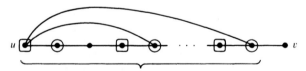

δu of these $n - 1$ vertices are circled

$\therefore$ δu of these $n - 1$ vertices are squared

$\therefore$ $n - \delta u - 1$ of these vertices are not squared

How many of these $n - 1$ vertices are joined to v?

$$\text{number of those vertices joined to } v = \delta v \geq n - \delta u > \text{number of those vertices which are not squared}$$

The first theorem of graph theory

It follows that v must be joined to some squared vertex:

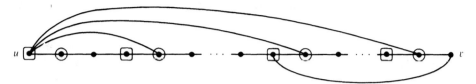

But then the graph G has the following Hamiltonian cycle, which contradicts the choice of G.

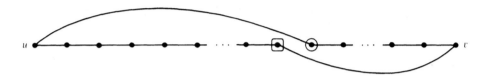

This contradiction shows that the theorem never fails. □

Exercises

1. (i) For which n is the complete graph K_n Eulerian and for which n is it semi-Eulerian? [A]
   (ii) For which m and n is the complete bipartite graph $K_{m,n}$ Eulerian and for which m and n is it semi-Eulerian? [A]
   (iii) For which n is K_n Hamiltonian? [A]
   (iv) For which m and n is $K_{m,n}$ Hamiltonian? [A]

2. Consider an $n \times n$ 'chess-board'. For what values of n is it possible to find a knight's route around the board which uses every possible *move* just once (in one direction or the other)? [A]

3. Let G be a connected graph with $k > 0$ vertices of odd degree. Show that there exist $\frac{1}{2}k$ paths in G which between them use each edge of G exactly once. [H]

4. The *complement* of a graph $G = (V, E)$ is the graph $\bar{G} = (V, \bar{E})$ where
   $$\bar{E} = \{vw: v, w \in V \text{ with } v \neq w \text{ and } vw \notin E\}$$
   (i.e. $\bar{G}$ consists of all the edges not in G). Find an example of a graph G which is Eulerian and whose complement is also Eulerian. [A]

5. Let G be a graph with $2d + 1$ vertices each of degree d. Show that G is Eulerian. [H]
   (Very keen readers might like to consider the much more difficult problem of whether, for $d > 0$, the graph G is also Hamiltonian.)

6. Let $G = (V, E)$ be a graph with $|V| \geq 3$ and with $\delta v \geq \frac{1}{2}|V|$ for each $v \in V$. Show that G is Hamiltonian.

7. Let $G = (V, E)$ be a graph with $|V| \geq 4$ and with the property that for any three of its vertices u, v and w, at least two of the edges uv, uw and vw are in E. Show that G is Hamiltonian.

8. In a group of $2n$ schoolchildren each one has at least n friends. On an outing the teacher tells them to hold hands in pairs. Show that this can be done with each person holding a friend's hand and that, if $n > 1$, the choice of friends can be made in at least two different ways.

9. There are n different examinations, each set by a single examiner, and each examiner sets at most half the papers. Show that the exams can be timetabled for one a day over a period of n consecutive days so that no examiner has two papers on consecutive days and so that the paper with the most candidates can be timetabled for the first day. [H]

10. Show that if a graph has n vertices and more than $\frac{1}{2}(n^2 - 3n + 4)$ edges then it is Hamiltonian. Show also that the corresponding statement with 'more than' replaced by 'at least' is false for each $n > 1$. [H]

11. A graph $G = (V, E)$ is *semi-Hamiltonian* if it possesses a path which uses each vertex of the graph exactly once. Show that, if $\delta u + \delta v \geq |V| - 1$ for each unjoined pair of vertices u and v, then G is semi-Hamiltonian (it may, of course, also be Hamiltonian). [H]

12. If $G = (V, E)$ is a graph and $v_1, \ldots, v_r$ are some (but not all) of the vertices of G then $G \backslash \{v_1, \ldots, v_r\}$ denotes the graph obtained from G by removing the vertices $v_1, \ldots, v_r$ (together with any edges ending at any of them).

    (i) Show that, if $|V| \geq 3$ and if either G consists of a single cycle or G is a complete graph, then $G \backslash \{u\}$ is connected for each $u \in V$ but $G \backslash \{u, v\}$ is disconnected for each pair $u, v \in V$ with $uv \notin E$.

    (ii) Conversely show that, if $|V| \geq 3$ and if $G \backslash \{u\}$ is connected for each $u \in V$ but $G \backslash \{u, v\}$ is disconnected for each pair $u, v \in V$ with $uv \notin E$, then either G consists of a single cycle or G is a complete graph.

    (The converse – which is used in chapter 11 – is much harder: in fact you only need the conditions that $G \backslash \{u\}$ is always connected and that $G \backslash \{u, v\}$ is disconnected for those pairs of vertices u, v with $uv \notin E$ for which there exists a vertex w with $uw, wv \in E$.) [H]

7
Edge-colourings

Given a graph, we wish to colour its edges so that no two edges of the same colour have a common endpoint (naturally enough called an *edge-colouring*). For any particular graph what is the smallest number of colours needed?

Examples

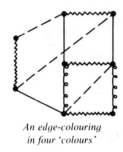

An edge-colouring in four 'colours'

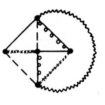

An edge-colouring in five 'colours'

In each of these examples the highest degree of any vertex is 4 and so it is clear that at least four colours are needed in any edge-colouring of either graph. In the left-hand example four colours are sufficient, but in the right-hand example a fifth colour is needed. □

For a graph G the minimum number of colours needed in an edge-colouring of it is called the *edge-chromatic number* of G and is denoted by $\chi_e(G)$. It is clear that if G has highest vertex degree d then $\chi_e(G) \geq d$, and for some graphs (like the left-hand example above) those d colours are actually sufficient and we have $\chi_e(G) = d$. Our first simple theorem provides a class of graphs for which $\chi_e(G) > d$.

Theorem *Let G be a graph with an odd number of vertices, each of which has positive degree d. Then G cannot be edge-coloured in d colours.*

Proof Let G have n vertices, where n is odd, each of them being of degree $d > 0$. Then in an edge-colouring of G no colour can occur more than once at each vertex and so there can be at most $\tfrac{1}{2}n$ edges of any one colour. So as n is odd there can be at most $\tfrac{1}{2}(n-1)$ edges of each colour. But as each vertex of the graph has degree d there are $\tfrac{1}{2}dn$ edges altogether. Therefore the number of different colours needed in any edge-colouring is at least

$$\frac{\tfrac{1}{2}dn}{\tfrac{1}{2}(n-1)} = d\,\frac{n}{n-1} > d;$$

and more than d colours are required. □

This result helps us to find the edge-chromatic numbers of the complete graphs.

Examples

$\chi_e(K_4) = 3$

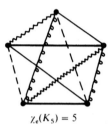

$\chi_e(K_5) = 5$ □

Theorem *The edge-chromatic number of the complete graph K_n is given by*

$$\chi_e(K_n) = \begin{cases} n-1 & \text{if } n \text{ is even,} \\ n & \text{if } n \text{ is odd.} \end{cases}$$

Proof Each vertex of K_n has degree $n-1$ and so it is immediate from the previous theorem that, when n is odd, more than $n-1$ colours are needed to edge-colour it. We shall show that $\chi_e(K_n) = n$ in this case by describing how to edge-colour K_n with n colours. Let the n vertices form a regular n-sided polygon and draw each edge as a straight line. Then simply colour two edges the same colour if and only if they are parallel. (The resulting colouring in the case $n = 5$ is shown in the right-hand example above and the case $n = 7$ is illustrated on the left at the top of the next page.) For n odd this will always give an edge-colouring of K_n in n colours; for it is clear that n colours are used, that no two edges of the same colour meet, and that all $\tfrac{1}{2}n(n-1)$ edges get coloured (each colour being used for exactly $\tfrac{1}{2}(n-1)$ edges). Hence the result is established for odd n.

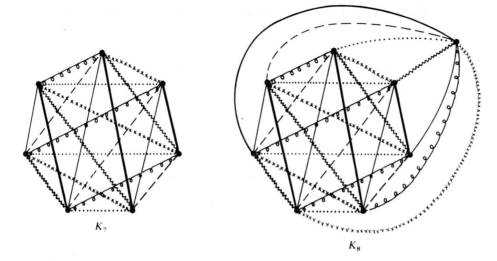

K_7 K_8

Now let n be even and edge-colour K_{n-1} with $n-1$ colours as described above. Then, in this colouring, there is one colour missing from each vertex, and furthermore each of the $n-1$ colours is missing from just one of the vertices. Add an nth vertex and join it to each of the other $n-1$ vertices in turn, each time using the colour missing from that vertex. It is easy to see that this gives an edge-colouring of K_n in $n-1$ colours: the case $n=8$ is illustrated on the right above. □

So each complete graph with an even number of vertices has the property that its edge-chromatic number equals its highest vertex degree and we are about to exhibit another class of graphs with the same property, namely the bipartite graphs. Before proceeding to that result we mention a trick which we shall use in the proofs of the remaining theorems in this chapter. If you are given an edge-colouring of a graph $G = (V, E)$ and you look at the subgraph on vertex-set V formed by the edges coloured in either of two colours, c and c' say, then in that subgraph each vertex will have degree 0, 1 or 2. So each non-trivial component of that subgraph will consist either of a cycle, or of a path with no repeated vertex, in which the colours of the edges alternate c/c'. A little thought will show you that reversing the colours throughout any component (i.e. using the colour c for edges which were formerly coloured c', and vice versa) will still leave an acceptable edge-colouring of G in the same colours, as illustrated in the following example.

Example

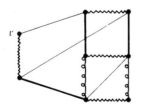

reverse the colours in the component containing v

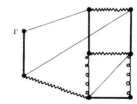 □

Theorem (König) *Let $G = (V_1, E, V_2)$ be a bipartite graph with highest vertex degree d. Then $\chi_e(G) = d$.*

Proof The proof is by induction on $|E|$, the case $|E| = 0$ being trivial. So assume that $G = (V_1, E, V_2)$ has $|E| > 0$, has highest vertex degree d, and that the result is known for bipartite graphs of fewer edges. Remove an edge vw from G to give an new bipartite graph G'. The highest vertex degree in G' is d or less and so by the induction hypothesis we can edge-colour G' in d colours: carry out such a colouring.

If one of the d colours is missing from both v and w then that colour can clearly be used for the edge vw to give the required edge-colouring of G in d colours.

On the other hand, assume that there is no colour missing from both v and w. Since the degrees of v and w in G' are at most $d - 1$ it follows that there is a colour, c say, which is missing from v (and hence is used at w) and there is another colour, c' say, which is missing from w (and hence is used at v). Now consider the subgraph of G' formed by the edges coloured in c or c', and in particular consider the component of this subgraph which contains v ($\in V_1$ say). This component consists of a path with edges coloured c' then c then $c',\ldots,$ starting at $v \in V_1$. This path cannot end at $w \in V_2$ for that would involve the path having an odd number of edges and hence ending with an edge of colour c', but no such edge ends at w. Now, by our earlier comments, we can reverse the two colours in this component and still have an acceptable edge-colouring of G' in d colours:
e.g.

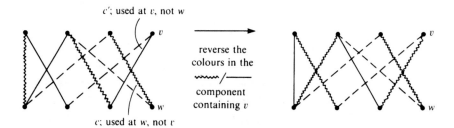

The advantage of this colour change is that the colour c' is now missing from both v and w: hence that colour can be used for the edge vw to give an edge-colouring of G in d colours, as required. This completes the proof by induction. □

So we have seen some graphs (such as the bipartite graphs and the complete graphs on an even number of vertices) for which the edge-chromatic number coincides with the highest vertex degree, d. We have also seen examples of graphs where d colours are not enough. But in those cases how many extra colours *will* be needed? It turns out that just one more colour will always be enough! Amazingly, no matter which graph you consider, if it has highest vertex degree d, then it can be edge-coloured in just $d + 1$ colours. This beautiful result was first proved by the Russian V.G. Vizing in 1964 and our proof of it uses the following lemma.

Edge-colourings

Lemma *Let $G = (V, E)$ be a graph with highest vertex degree d and let $e_1, \ldots, e_r$ be some edges of G with a common endpoint v. Assume that $G' = (V, E \setminus \{e_1, \ldots, e_r\})$ can be edge-coloured in D ($\geq d$) colours in such a way that there is at least one of the colours not occurring at either end of e_1, there are at least two colours not occurring at either end of e_2, there are at least two colours not occurring at either end of $e_3, \ldots$ and there are at least two colours not occurring at either end of e_r. Then G can be edge-coloured in D colours.*

Proof The proof is by induction on r. The case $r = 1$ is trivial since, if G' is coloured in the stated way, then we are given that there is at least one colour not used at either end of the single missing edge e_1: clearly that colour can be used for e_1 to complete the edge-colouring of G.

So now assume that $r > 1$, that the result is known for fewer than r removed edges, and that we are given an edge-colouring of G' as stated. We shall show how to colour one of the r edges, but we must do it in such a way that the other $r - 1$ uncoloured edges still satisfy the conditions of the lemma. Consider the following given sets of some of the colours not occurring at the ends of the e_i:

$$C_1 = \{\text{a colour not occurring at either end of } e_1\},$$
$$C_2 = \{\text{two colours not occurring at either end of } e_2\},$$
$$C_3 = \{\text{two colours not occurring at either end of } e_3\},$$
$$\vdots$$
$$C_r = \{\text{two colours not occurring at either end of } e_r\}.$$

If some colour occurs in just one of these sets then it is possible to use that colour for that edge, to delete that set from the list and to leave a new situation where G has only $r - 1$ uncoloured edges and where they still satisfy the conditions of the lemma. Then applying the induction hypothesis to this new situation shows us that G itself can be edge-coloured in the D colours.

But what if there is no colour which occurs in just one of the sets? Then each colour in one of the sets occurs in at least two of the sets: it follows that the number of different colours in all the sets is at most $\frac{1}{2}(2r - 1)$ and, in particular, is less than r. In addition we know that the vertex v has degree at most d in G and so in G' there are at most $d - r$ colours occurring at v. Therefore

$$\begin{array}{c}\text{number of different}\\ \text{colours not}\\ \text{occurring at } v\end{array} \geq D - (d - r) \geq d - (d - r) > \begin{array}{c}\text{number of different}\\ \text{colours in the } r \text{ sets}\\ C_1, \ldots, C_r.\end{array}$$

Hence there is a colour, c say, not occurring at v and not in any of the given sets of colours. Let $C_1 = \{c'\}$ and look at the subgraph formed by the edges coloured c or c' in the given colouring of G'. As commented earlier we can reverse the two colours in any component of this subgraph and still leave an acceptable colouring of G'. In particular assume that the edge e_1 joins v to v_1 and reverse the two colours c and c' in the component which contains v_1. This recolouring will mean that the colour c is now not used at either end of e_1, so then use the colour c for the edge e_1.

e.g.

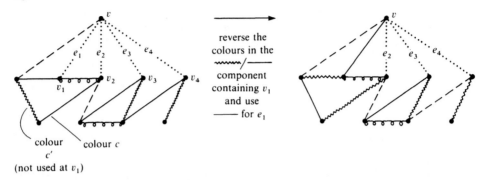

Then delete C_1 from the list of sets and revise the sets $C_2, \ldots, C_r$ in line with this new colouring. What form will this revision take? The only necessary change will be if one of the sets (C_2 say) contains c' and the colour c' is now used at the end v_2 of the edge $e_2 = vv_2$ (which happens if v_1 and v_2 are the two 'ends' of the same c/c' component, as in the case illustrated above). In that event we must delete c' from the set C_2. In all other cases we will do nothing.

In each case we are now left with all but $r - 1$ edges of G coloured, those edges still having the common endpoint v, and the first of these uncoloured edges has a set of at least one colour not used at either of its ends and each of the other uncoloured edges has a set of two given colours not used at either of its ends. We can now apply the induction hypothesis to this situation and deduce that the whole of G can be edge-coloured in the D colours, and the lemma is proved. □

Example *The following graph has been partially edge-coloured in the five colours 1–5 and the uncoloured edges satisfy the conditions of the lemma. Use the technique of the above proof to adapt the edge-colouring to one for the whole of the graph.*

Solution

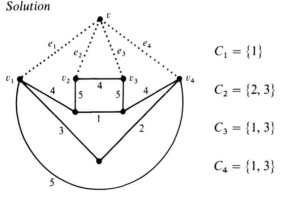

$C_1 = \{1\}$ (a colour not used at either end of e_1).

$C_2 = \{2, 3\}$ (two colours not used at either end of e_2).

$C_3 = \{1, 3\}$ (two colours not used at either end of e_3).

$C_4 = \{1, 3\}$ (two colours not used at either end of e_4).

Is there a colour appearing in just one of those sets? Yes, the colour 2 only appears in the set C_2. Use that colour for the edge e_2 and start again:

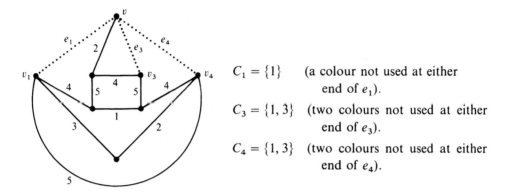

$C_1 = \{1\}$ (a colour not used at either end of e_1).

$C_3 = \{1, 3\}$ (two colours not used at either end of e_3).

$C_4 = \{1, 3\}$ (two colours not used at either end of e_4).

Is there a colour appearing in just one of the sets? No. So choose a colour not used at v and not appearing in any of the sets (we have seen that there will always be one): in this case colour 4 will do. Take as a second colour the one in the first set, namely 1. Then look at the subgraph formed by edges of colour 1 or 4. In the component of that subgraph which contains v_1 reverse the two colours 1 and 4. As we saw earlier this will still leave an acceptable colouring of the same set of edges and it will free the colour 4 to be used for the first edge e_1. But in this case it leaves the colour 1 used at v_4 and so we must delete 1 from C_4 and start again. (We also move the revised C_4 to first place in the list as it is now the singleton set.)

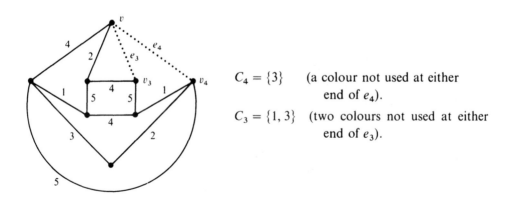

$C_4 = \{3\}$ (a colour not used at either end of e_4).

$C_3 = \{1, 3\}$ (two colours not used at either end of e_3).

Now colour 1 only occurs in C_3 so that colour can be used for e_3, and the remaining colour 3 can be used for the other edge e_4:

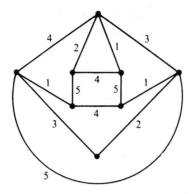

□

We are now ready to prove Vizing's remarkable theorem.

Theorem (Vizing) *Let $G = (V, E)$ be a graph with highest vertex degree d. Then G can be edge-coloured in $d + 1$ colours; i.e. $d \leq \chi_e(G) \leq d + 1$.*

Proof The proof is by induction on $|E|$, the case $|E| = 0$ being trivial. So assume that $|E| > 0$ and that result is known for graphs of fewer edges. Let $v \in V$ have positive degree and be the endpoint of precisely the edges $e_1 = vv_1, \ldots, e_r = vv_r$, and let G' be the graph $(V, E \setminus \{e_1, \ldots, e_r\})$. Then the highest degree in G' is d or less and so by the induction hypothesis the graph G' can be edge-coloured in $d + 1$ colours: carry out such a colouring.

In G' the degree of each of the vertices $v_1, \ldots, v_r$ is less than or equal to $d - 1$ and there are $d + 1$ colours available. So for each v_i there are at least two colours not used there and hence not used at either end of e_i. In particular there is one such colour for e_1 and two colours for each of the others. Therefore by the lemma (with $D = d + 1$) there is an edge-colouring for the whole of G in those $d + 1$ colours, and the proof by induction is complete. □

We now see an extension of Vizing's theorem which gives a sufficient (but not necessary) condition for a graph to be edge-colourable in d colours.

Theorem *Let $G = (V, E)$ be a graph with highest vertex degree d (≥ 2) and such that no cycle of G consists entirely of vertices of degree d. Then G can be edge-coloured in d colours; i.e. $\chi_e(G) = d$.*

Proof Surprisingly, the proof of Vizing's theorem which we have used can be adapted almost immediately to prove this extension. Again the proof is by induction on $|E|$, the cases $|E| \leq 2$ being trivial. So assume that $|E| > 2$ and that the result is known for graphs of fewer edges. Since no cycle in G consists entirely of vertices of degree d it follows that there is a vertex of degree d which is joined to at most one other vertex of degree d. Let $v \in V$ be such a vertex, assume that it is the endpoint of the edges $e_1 = vv_1, \ldots, e_d = vv_d$, and let G' be the graph $(V, E \setminus \{e_1, \ldots, e_d\})$. We show first that G' can be edge-coloured in d colours.

The highest vertex degree in G' is d or less: if it is less than d then Vizing's theorem itself tells us that G' can be edge-coloured in d colours. If, on the other hand, the highest degree in G' is d then, as it is still true that no cycle in G' consists entirely of vertices of degree d, it follows that we can apply the induction hypothesis to the graph G' to show once again that it is edge-colourable in d colours. So in either case edge-colour G' in d colours.

The fact that v was joined to at most one vertex of degree d in G means that at most one of the vertices $v_1, \ldots, v_d$ has degree d in G and the rest have degree $d-1$ or less: if one of these vertices does have degree d in G then assume that it is v_1. Hence in G', from the vertices $v_1, \ldots, v_d$, at most v_1 has degree $d-1$ and the rest have degree less than or equal to $d-2$. As there are d colours available in our edge-colouring of G', it follows that there is at least one colour not used at v_1 (and hence not used at either end of the edge e_1) and for each of the other v_i there are at least two colours not used there (and hence not used at either end of the corresponding e_i). Therefore by the lemma (with $D = d$) there is an edge-colouring for the whole of G in d colours, and the proof by induction is complete. □

The proof of Vizing's theorem (and of its extension) proceeded by adding on a vertex at a time and then using the process described in the proof of the lemma to colour the additional edges at each stage. So, as illustrated in the previous example, given a graph of highest vertex degree d these proofs can actually be used to give an algorithm for edge-colouring the graph in $d+1$ (or in d) colours. The process is described further in *Matching theory* by L. Lovasz and M.D. Plummer, and other proofs of Vizing's theorem, together with a selection of related results, can be found in the graph theory books by M. Behzad and C. Chartrand, by S. Fiorini and R.J. Wilson, and by L.W. Beineke and R.J. Wilson cited in the bibliography. In that last book those graphs for which the edge-chromatic number coincides with the highest vertex degree are said to be of 'class 1' and the rest are of 'class 2': there is no known simple characterisation of the two classes although the authors discuss the extent to which the class 2 graphs are 'quite rare'.

We shall return to the topic of edge-colouring in chapter 14, where we see a very neat connection between edge-colouring and map-colouring.

Exercises

1. Show that, for any graph $G = (V, E)$,

$$\chi_e(G) \geq \frac{|E|}{[\frac{1}{2}|V|]},$$

   where $[x]$ denotes the integer part of x.

2. Let G be a graph in which each vertex except one has degree d. Show that if G can be edge-coloured in d colours then
   (i) G has an odd number of vertices, [H]
   (ii) G has a vertex of degree zero. [H]

3. Let G be a connected graph in which each vertex has degree d and which has a vertex whose removal (together with all the edges ending there) disconnects G. Show that $\chi_e(G) = d + 1$. [H]

4. Let G be a Hamiltonian graph in which each vertex has degree 3. Show that $\chi_e(G) = 3$. [H]

5. In a class each boy knows precisely d girls and each girl knows precisely d boys. Use a result on edge-colouring to show that the boys and girls can be paired off in friendly pairs in at least d different ways. [H]

6. Let M be an $m \times n$ matrix of 0s and 1s with at most d 1s in any row or column. Show that M can be expressed as the sum of d matrices of 0s and 1s each of which has at most one 1 in any row or column. [H]

8
Harems and tournaments

We begin this short chapter by looking at an extension of Hall's marriage theorem to harems, where some people are allowed more than one partner. We then consider a different sort of match by looking at 'round robin tournaments' in a sports club. The unexpected connection between the two worlds comes when we are able to use our harem result to deduce a theorem about tournaments.

In chapter 3 we met Hall's marriage theorem and now we are going to deduce a slight generalisation of it. Traditionally in a harem each man can have many wives but no woman can have more than one husband. We now deduce the harem version of Hall's theorem, but we do our small bit to put right the sexual imbalance.

Theorem (Hall's theorem – harem form) *Let $b_1, \ldots, b_n$ be non-negative integers and let $G_1, \ldots, G_n$ be girls. Girl G_1 wants b_1 husbands (as always, from amongst the boys whom she knows), girl G_2 wants b_2 husbands, $\ldots$, and girl G_n wants b_n husbands. No boy can marry more than one girl. Then the girls' demands can all be satisfied if and only if any collection of the girls $G_{i_1}, \ldots, G_{i_s}$ knows between them at least $b_{i_1} + \cdots + b_{i_s}$ boys.*

Proof ($\Rightarrow$) If the girls can all find the required numbers of husbands, then the girls $G_{i_1}, \ldots, G_{i_s}$ (as always, assumed to be distinct) must know between them at least $b_{i_1} + \cdots + b_{i_s}$ boys, namely their husbands.

($\Leftarrow$) Assume that any collection $G_{i_1}, \ldots, G_{i_s}$ of girls know between them at least $b_{i_1}, \ldots, b_{i_s}$ boys. Replace G_1 by b_1 copies of G_1 each of whom knows the same boys that G_1 did (so that, for example, if $b_1 = 3$ then G_1 is replaced by triplets). Similarly, replace G_2 by b_2 copies of $G_2, \ldots$ and G_n by b_n copies of G_n, giving:

$$\text{girls:} \quad \underbrace{G_1 \cdots G_1}_{b_1} \quad \underbrace{G_2 \cdots G_2}_{b_2} \quad \cdots \quad \underbrace{G_n \cdots G_n}_{b_n}$$

We shall apply Hall's marriage theorem to find each girl one husband in this new situation. So we shall show that, in this new situation, any r girls know between them at least r boys. Choose any set of these new girls: let there be r of them, say, and assume these these r consist of at least one 'copy' of each of the girls $G_{i_1}, \ldots, G_{i_s}$. Then clearly

$$r \leqslant b_{i_1} + \cdots + b_{i_s}$$

since the right-hand total is the number we would get if we took *all* copies of those girls. Now the given conditions tell us that in the original situation the girls $G_{i_1}, \ldots, G_{i_s}$ knew between them at least $b_{i_1} + \cdots + b_{i_s}$ boys (and hence at least r boys). It follows that our set of r girls (which includes a copy of each of $G_{i_1}, \ldots, G_{i_s}$) in the new twin/triplet situation know between them at least r boys.

Hence, as any r girls do know at least r boys, we can apply Hall's marriage theorem to the new situation to find each of the girls a husband. So, for example, each of the b_1 copies of G_1 will get a husband: clearly these can be combined to give b_1 different husbands for the original girl G_1. Proceeding in this way gives each girl the required number of husbands in the original situation, and the harem theorem is proved. □

We now turn to the world of tournaments. In a squash club, for example, a 'round robin tournament' for a group of players consists of each of them playing each of the others once, each game resulting in a win for one of the two players. Imitating this idea, a *tournament* of n players (called $1, 2, \ldots$ and n, say) consists of $\binom{n}{2}$ ordered pairs of players so that for $1 \leq i < j \leq n$ either the pair ij or the pair ji is included. The $\binom{n}{2}$ ordered pairs can be thought of as the results of the games between each pair of players, with the first number in the pair being the winner of that game: in this way we shall use all the normal sports jargon. In addition a tournament of n players can be represented by a picture of the complete graph K_n with the vertices labelled $1, 2, \ldots, n$ and with an arrow on each edge, an arrow from i to j meaning that 'i beat j'.

Example *The diagram on the right represents a tournament of the five players 1, 2, 3, 4 and 5 with pairs* 12, 13, 23, 24, 34, 41, 51, 52, 53 *and* 54. *So in this tournament* 1 *beat* 2, 1 *beat* 3, 2 *beat* 3, 2 *beat* 4, 3 *beat* 4, 4 *beat* 1, *and* 5 *beat everyone.*

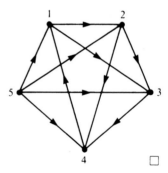

□

The graphical representation of a tournament gives a special sort of 'directed graph', a topic which we do not cover in general in this book. A *directed graph* (or *digraph*) $(V, \vec{E})$ consists of a set of vertices, as before, and a set of edges $\vec{E}$ which is a subset of the *ordered* pairs $\{vw : v, w \in V\}$. A directed graph can be illustrated in the same way that we illustrated graphs, but with arrows on edges to indicate the order. The terminology of graph theory extends naturally to directed graphs: for example a 'directed path' is merely a path followed in the directions of the arrows, and the 'out-degree'/'in-degree' of a vertex is the number of edges which start/end at that vertex, etc. In the graphical representation of a tournament a directed path $p_1, p_2, \ldots, p_r$ is a sequence of players in which p_1 beat p_2, p_2 beat $p_3, \ldots$ and p_{r-1} beat p_r. For instance, in the above example there is a directed path 5, 4, 1, 2, 3 which uses all the vertices. Imitating our work on Hamiltonian and semi-Hamiltonian graphs in chapter 6 we now show that there always exists a directed path in a tournament which uses each player exactly once: then in the exercises we find conditions for there to exist a directed Hamiltonian cycle.

Theorem *In a tournament of n players they can be labelled $p_1, p_2, \ldots, p_n$ so that p_1 beat p_2, p_2 beat $p_3, \ldots$ and p_{n-1} beat p_n.*

Proof The proof is by induction on n, the case $n = 2$ (or $n = 1$) being trivial. So assume that $n > 2$ and that the result is known for tournaments of fewer than n players. For the moment forget one of the players, p say, and consider the games between the other $n - 1$ players. These will form a tournament and therefore, by the induction hypothesis, the remaining $n - 1$ players can be labelled $p_1, p_2, \ldots, p_{n-1}$ so that

$$p_1 \text{ beat } p_2, p_2 \text{ beat } p_3, \ldots, p_{n-2} \text{ beat } p_{n-1}.$$

Now think about the results of p's games. If p was beaten by all the other players then we will have

$$p_1 \text{ beat } p_2, p_2 \text{ beat } p_3, \ldots, p_{n-2} \text{ beat } p_{n-1}, p_{n-1} \text{ beat } p$$

which is of the required form involving all n players in the tournament. If on the other hand p beat at least one of the other players, then let i be the *lowest* for which p beat p_i. Then if $i = 1$ we have

$$p \text{ beat } p_1, p_1 \text{ beat } p_2, p_2 \text{ beat } p_3, \ldots, p_{n-2} \text{ beat } p_{n-1}$$

and if $i > 1$ we have

$$p_1 \text{ beat } p_2, \ldots, \underbrace{p_{i-1} \text{ beat } p}_{\text{since } p_i \text{ was the first beaten by } p}, p \text{ beat } p_i, \ldots, p_{n-2} \text{ beat } p_{n-1}.$$

In both cases we have the required ordering of all the players, and the theorem is proved. □

A lot of results for graphs extend naturally to directed graphs: for example a directed graph has a directed Eulerian path if and only if it is 'strongly connected' (in the sense that there is a directed path from any one vertex to any other) and the out-degree of each vertex equals its in-degree. The interested reader can find out more about directed graphs in, for example, R.J. Wilson's book *Introduction to graph theory* and more about directed Hamiltonian cycles in J.-C. Bermond's chapter in the collection *Selected topics in graph theory*: tournaments are also covered in both these works and in J.W. Moon's *Topics on tournaments*. All these books are listed in the bibliography.

The main result about tournaments which is going to concern us here is about the possible collection of 'scores'. In a tournament of the n players $1, 2, \ldots, n$ let b_i be the number of players beaten by player i: then $b_1, b_2, \ldots, b_n$ are the *scores* of the tournament (and the ordered list of them is the *score vector*).

Example *Which of the following are possible scores in a tournament of six players?*
  (i) 4, 4, 4, 2, 1, 1;
  (ii) 5, 3, 3, 2, 1, 1;
  (iii) 5, 4, 4, 1, 1, 0.

Solution (i) Clearly these cannot be the scores in a tournament of six players because
$$4 + 4 + 4 + 2 + 1 + 1 = 16 \neq \binom{6}{2}.$$

(ii) Here (in graphical form) is a tournament with the given scores. Note that the scores correspond to the out-degrees in the graphical representation.

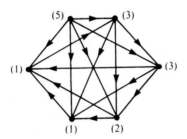

(iii) Although the six numbers 5, 4, 4, 1, 1, 0 do add to $\binom{6}{2}$ they cannot be the scores in a tournament of six players. One way of seeing this is to consider the games between the three players who are supposed to have scores of 1, 1 and 0. Those three players will have played $\binom{3}{2} = 3$ games amongst themselves and so, no matter how bad they are, those 3 wins must have been shared between them. Therefore it is impossible for their scores to add up to less than 3. □

The properties highlighted by those examples will be shared by the scores of any tournament of n players, namely the scores will add up to $\binom{n}{2}$, and any r of the scores will add up to at least $\binom{r}{2}$ (because those r players will have played that number of games just amongst themselves). Remarkably the converse also turns out to be true; i.e. if a collection of n integers has those two properties then they are the scores of a tournament. This result was first proved at length by H.G. Landau in 1953 in connection with his work on animal societies: it is only very recently that Landau's theorem has been recognised as an easy consequence of the harem version of Hall's theorem.

Theorem (Landau) *Let $b_1, b_2, \ldots, b_n$ be integers. Then they are the scores in a tournament of n players if and only if*

(i) $$b_1 + b_2 + \cdots + b_n = \binom{n}{2},$$

and

(ii) *for $1 \leq r \leq n$ any r of the b_i's add up to at least $\binom{r}{2}$.*

Proof ($\Rightarrow$) Assume that there exists a tournament of n players with scores $b_1, b_2, \ldots, b_n$. Then clearly the sum of these scores is the total number of games played, which is $\binom{n}{2}$. Also any r of the players will have played $\binom{r}{2}$ games amongst themselves and so they must have at least $\binom{r}{2}$ wins between them. Hence any r of the scores must add up to at least $\binom{r}{2}$. Thus properties (i) and (ii) are established.

($\Leftarrow$) Now let $b_1, b_2, \ldots, b_n$ be integers satisfying (i) and (ii): then, in particular, taking $r = 1$ shows that these integers are non-negative. Now given any s of the numbers, $b_{i_1}, b_{i_2} \ldots, b_{i_s}$ say, consider the *other* $n - s$ numbers: they will add up to at least $\binom{n-s}{2}$ and so it follows that

$$b_{i_1} + b_{i_2} + \cdots + b_{i_s} \leq \binom{n}{2} - \binom{n-s}{2}.$$

Now imagine that young ladies $1, 2, \ldots, n$ enter a squash tournament. Offer as prizes $\binom{n}{2}$ handsome young men called '1,2', '1,3', ..., '1,n', '2,3', '2,4', ... and '$n-1, n$': the boy 'i,j' is to be awarded as the prize in the game between players i and j. The girls get to know just the boys who they are playing for, so girl 2 for example gets to know the boys '1,2', '2,3', '2,4', ... and '2,n'. Give any s of the girls, $i_1, i_2, \ldots, i_s$, how many boys do they now know between them? They know all the boys except those exclusively known by the other $n - s$ girls; i.e. they know exactly

$\binom{n}{2} - \binom{n-s}{2}$ boys between them. So, by the inequality displayed above, the s girls $i_1, i_2, \ldots, i_s$ know between them at least $b_{i_1} + b_{i_2} + \cdots + b_{i_s}$ boys. Hence by the harem form of Hall's theorem, proved earlier, the girls $1, 2, \ldots, n$ can find $b_1, b_2, \ldots, b_n$ husbands, respectively, from amongst the boys they know. In other words the decisions about whether girl i or girl j should marry boy 'i,j' can be made so that girl 1 ends up with b_1 prizes, girl 2 with b_2 prizes, ... and girl n with b_n prizes. It follows that the results of all the games in the tournament can be fixed so that the number of prizes won by $1, 2, \ldots, n$ are $b_1, b_2, \ldots, b_n$ respectively: these numbers are then the scores of the tournament, and the theorem is proved. □

Exercises

1. In a tournament of n players,
   (i) how many different sets of results are there between the players so that no two of them have the same score? [A]
   (ii) how many different sets of results are there between the players so that two of them have the same score but, apart from that, no score is repeated? [A]
   (iii) if one player has the highest score and all the rest tie second, what are their scores? [A]

2. In a tournament of n players with scores $b_1, b_2, \ldots, b_n$ how many sets of three players are there with the property that each player in the three won one of the three games played amongst the three? [H,A]

3. Show that in a tournament of n players the sum of any r scores is at most
$$nr - \binom{r+1}{2}.$$
[H]

4. A tournament (or directed graph in general) is *strongly connected* if there is a directed path from any one vertex to any other. Show that for a tournament of n players the following three properties are equivalent:
   (i) the tournament is strongly connected;
   (ii) it has a directed Hamiltonian cycle;
   (iii) for $1 \leq r < n$ the sum of any r scores is greater than $\binom{r}{2}$. [H]

5. Show that in a tournament of n players there will be no ordered cycle if and only if the players can be labelled $p_1, \ldots, p_n$ such that p_i beat p_j whenever $i < j$ (i.e. so that p_1 beat everybody else, p_2 beat everybody except p_1, etc.). Deduce that there are $n!$ ways of choosing the results of a tournament of n players so that there is no ordered cycle. (Any such tournament of n players is called an 'acyclic orientation' of K_n, and we shall meet these again in chapter 11.)

6. (i) A tournament of n players takes place over a period of consecutive days and

no player is allowed to play more than one game a day. What is the minimum number of days required to complete the tournament? [H,A]

(ii) A football league consists of n teams and in the course of the season each team will play each of the others twice. No team plays more than once on any day. What is the minimum number of playing days needed in the season? [A]

9
Minimax theorems

Imagine that you have a bag of mixed sweets of several different types. What is the minimum number of sweets that you must choose in order to show me how many different types you have? What is the maximum number of sweets that you could take from the bag without getting two of the same type? The answers are, of course, the same. That is a trivial example of situation where one question concerning a minimum leads to the same answer as another question concerning a maximum: it is a 'minimax' result. In this chapter we look at three non-trivial minimax theorems; one concerns matrices, one is in graph theory, and the other is about networks.

Matrices

Here once again we are going to use the ubiquitous Hall's theorem. In its matrix form it concerns taking a matrix of 0s and 1s and looking for one 1 in each row with no two in the same column. However, such a set of 1s may not exist and we may instead look for as many 1s as possible with no two in the same row or column.

Examples

$$M_1 = \begin{pmatrix} 0 & 1 & 0 & 1 & 0 & 1 \\ 1 & 0 & 1 & 0 & 0 & 0 \\ 0 & 0 & 1 & 1 & 0 & 1 \\ 1 & 1 & 0 & 0 & 1 & 0 \\ 0 & 0 & 1 & 1 & 1 & 0 \end{pmatrix},$$

$$M_2 = \begin{pmatrix} 0 & 1 & 1 & 0 & 0 & 0 \\ 1 & 0 & 0 & 1 & 0 & 1 \\ 0 & 1 & 0 & 0 & 1 & 1 \\ 0 & 1 & 1 & 0 & 0 & 0 \\ 0 & 0 & 1 & 0 & 0 & 0 \end{pmatrix} = \begin{pmatrix} 0 & \boxed{1} & \boxed{1} & 0 & 0 & 0 \\ \boxed{1} & 0 & 0 & 1 & 0 & 1 \\ \boxed{0} & \boxed{1} & 0 & 0 & 1 & 1 \\ 0 & 1 & 1 & 0 & 0 & 0 \\ 0 & \boxed{0} & \boxed{1} & 0 & 0 & 0 \end{pmatrix}.$$

The matrix M_1 satisfies the Hall-type condition that in any r rows there are 1s in at least r columns. It therefore has a set of 1s with one in each row and no two in the same column, and one such set is shown in bold print. The matrix M_2, however, has no such set of 1s: it is only possible to find a maximum of *four* 1s with no two in the same row or column, and again one such collection is shown in bold print. Finally, note that it is possible to find *four* 'lines' (i.e. rows or columns) in M_2 which between them include all its 1s, as shown by the boxes in the right-hand figure. In M_1 the five rows clearly contain all its 1s and there is no smaller set of 'lines' which do so. □

Theorem (König–Egerváry) *Let M be an $m \times n$ matrix of 0s and 1s and let the word 'line' denote a row or column of M. Then*

$$\frac{\text{the minimum number of lines}}{\text{containing all the 1s of } M} = \frac{\text{the maximum number of 1s}}{\text{with no two in the same line.}}$$

Proof Let α be the minimum number of lines of M which between them contain all its 1s and let β be the maximum number of 1s of M with no two in the same line. Since there are β 1s with no two in the same line it clearly takes at least β lines to include those 1s: hence $\alpha \geq \beta$.

To show that $\beta \geq \alpha$ we shall find α 1s with no two in the same line. Assume that $\alpha = p + q$ where (with no loss in generality) the first p rows and the first q columns of M between them contain all the 1s of M.

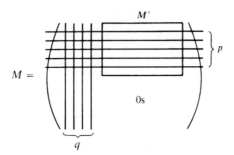

Look at the matrix M'. We claim that any r rows of it have between them 1s in at least r columns. For if not there would be some r rows of M' with 1s in just r' columns, where $r' < r$:

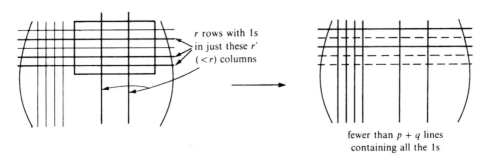

But then replacing the corresponding r rows of M by those r' columns would give a smaller set of lines containing all the 1s of M, which is not possible. Hence any r

rows of M' do contain 1s in at least r columns. Therefore by the matrix form of Hall's theorem (page 31) M' has a set of p 1s with one in each row and no two in the same column.

We have shown that there are p 1s in M' with no two in the same line. Similarly there are q 1s in M'' with no two in the same line. Clearly overall that gives us $p + q$ ($=\alpha$) 1s in M with no two in the same line. Then, since β is the maximum number of such 1s, it follows that $\alpha \leq \beta$, and the theorem is proved. □

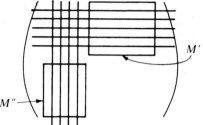

Graphs

In a graph a *matching* is a set of edges with no two endpoints in common. We have already met the idea of a 'matching from V_1 to V_2' in the bipartite graph $G = (V_1, E, V_2)$: if one exists then it is clear that G has no matching with a larger number of edges. However, a bipartite graph $G = (V_1, E, V_2)$ might not possess a matching from V_1 to V_2, in which case its largest matching will have fewer than $|V_1|$ edges.

Examples

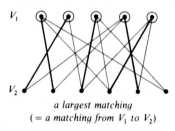

a largest matching
(= a matching from V_1 to V_2)

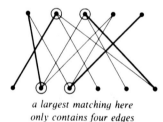

a largest matching here
only contains four edges

In the left-hand example the five vertices ringed (which form V_1) clearly have the property that every edge of the graph has at least one of them as an endpoint: no set of fewer than five vertices can be found with the same property. In the right-hand example the set of four ringed vertices is the smallest collection which between them include at least one endpoint of each edge of the graph. □

That brings us to our next minimax theorem. Eagle-eyed readers will have seen a strong connection between that example and the previous one on matrices, and in fact the following result is virtually a restatement of the König–Egerváry theorem in graph-theoretic terms: its proof is left as one of the exercises at the end of the chapter.

Theorem *Let $G = (V_1, E, V_2)$ be a bipartite graph. Then*

the minimum number of vertices which between them include at least one endpoint of each edge = *the maximum number of edges in a matching.* □

Both the above results are central to some applications which, although they are beyond the scope of this book, are worth a few comments so that the interested reader can look elsewhere. In *A first course in combinatorial mathematics* I. Anderson illustrates how the König–Egerváry theorem is the key to an algorithmic method for solving the 'optimal assignment problem' of allocating jobs to people in the best way possible. In D.R. Woodall's chapter in *Selected topics in graph theory* (which is an interesting survey of minimax theorems) he describes the *Hungarian algorithm* for finding a largest matching in a bipartite graph: the method (which is so called because of the nationalities of König and Egerváry) is not unlike the algorithmic proof of Hall's theorem which we used in chapter 3. Woodall also shows how this method extends to a 'weighted' version which then in turn can be used to solve the optimal assignment problem.

Another interpretation of the last theorem in a bipartite graph is that in some sense the maximum number of disjoint paths from V_1 to V_2 equals the minimum number of vertices whose removal disconnects V_1 from V_2. Remarkably the next theorem, first proved by K. Menger in 1927, shows that the same is true for any two sets of vertices in any graph.

A collection of paths in a graph $G = (V, E)$ is *disjoint* if no two of them have any vertex in common. A path from a set of vertices X to a set of vertices Y is simply a path whose first vertex is in X and whose last vertex is in Y. When we are looking for the largest possible collection of disjoint paths from X to Y we shall naturally look for the most economical ones and restrict attention to those paths whose first vertex is the *only* one in X and whose last vertex is the *only* one in Y. On the other hand, a set S of vertices *separates* X from Y if every path from X to Y uses a vertex from S: the sets X and Y themselves trivially have this property, but there are often smaller ones.

Example *In the graph illustrated there is a maximum of three disjoint paths from X to Y (one such set is shown in bold print). Also there is a minimum of three vertices whose removal separates X from Y (one such set is shown ringed).*

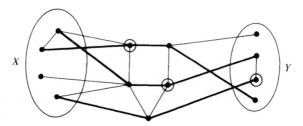

□

Theorem (Menger) *Let $G = (V, E)$ be a graph and let $X, Y \subseteq V$. Then*

$$\begin{array}{c} \textit{the minimum number of vertices} \\ \textit{which separate } X \textit{ from } Y \end{array} = \begin{array}{c} \textit{the maximum number of} \\ \textit{disjoint paths from } X \textit{ to } Y. \end{array}$$

Proof Let α be the minimum number of vertices which separate X from Y, and let β be the maximum number of disjoint paths from X to Y. As usual with these minimax theorems one inequality between α and β is obvious: if a set of vertices separates X from Y it must have at least one vertex on each of the β disjoint paths and so $\alpha \geqslant \beta$.

The proof of the reverse inequality is by induction on $|V| + |E|$, the case $|V| + |E| = 1$ being trivial. So assume that G has $|V| + |E| > 1$ and that the result is known for smaller graphs: to show that $\beta \geqslant \alpha$ for G we shall construct α disjoint paths from X to Y. We deal with the proof in two cases:

Case 1: There is a set S of α vertices which separates X from Y and with $S \neq X$ and $S \neq Y$. (Case 2 will be all other cases.) As Y itself separates X from Y, and S is a minimal such set, it follows that $Y \not\subset S$. As in this case we also know that $S \neq Y$ it follows that $Y \not\subseteq S$ and that there is some vertex $y \in Y \setminus S$. A little thought will soon show you that S is a smallest set of vertices which separates X from S, and not only is this true in G but it is also true in the graph obtained from G by deleting the vertex y.

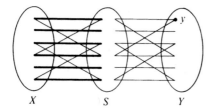 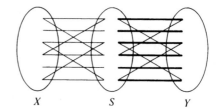

Hence, by the induction hypothesis, in this smaller graph there exist α disjoint paths from X to S, and these will still be disjoint paths in G. By a similar argument there exist α disjoint paths from S to Y in G. The paths from X to S can clearly be strung together with the paths from S to Y to give the required α disjoint paths from X to Y.

Case 2. If case 1 fails the only possible set of α vertices to separate X from Y will be one of X and Y themselves (or both): assume, say, that $|X| = \alpha$. If $X \subseteq Y$ then it is easy to construct α trivial disjoint paths from X to Y, each consisting of a single vertex. So assume that $v \in X \setminus Y$ and consider the set of vertices $X \setminus \{v\}$. Since this has fewer than α vertices it does not separate X from Y: therefore there exists a path from X to Y which begins $v, w, \ldots$. Now consider the graph $G' = (V, E \setminus \{vw\})$. If X is still a minimal separating set in G' then by the induction hypothesis there exist α disjoint paths from X to Y in G' (and hence in G). On the other hand, if there is some set S' of $\alpha - 1$ vertices which separates X from Y in G' then it is clear that the sets $S' \cup \{v\}$ and $S' \cup \{w\}$ both separate X from Y in G and that both contain α vertices. But in case 2 the only possible such sets are X and Y themselves. Therefore

$$S' \cup \{v\} = X \quad \text{and} \quad S' \cup \{w\} = Y.$$
$$\uparrow \qquad\qquad\qquad \uparrow$$
$$\in X \setminus Y \qquad\qquad \notin X$$

But then it is clear that the $\alpha - 1$ trivial paths consisting of a single vertex of S' together with the path v, w form α disjoint paths from X to Y as required.

Hence in all cases there exist α disjoint paths from X to Y: as β was the maximum number of such paths it follows that $\beta \geqslant \alpha$. This, together with the opposite inequality established earlier, shows that $\alpha = \beta$ and Menger's theorem is proved. □

Networks

A *network* $N = (V, c)$ is a finite non-empty set of vertices V such that with each ordered pair of vertices $v, w \in V$ there is associated a non-negative real number $c(vw)$ called the *capacity* of vw (i.e. c is a function from $V \times V$ to $\{x \in R: x \geqslant 0\}$.). Without getting bogged down in the terminology of direct graphs, if $c(vw) > 0$ then we can think of vw as an edge from v to w and illustrate a network in the same way that we illustrate a graph, noting the capacities of the edges but totally ignoring edges of zero capacity. Also, since we are not studying networks very extensively, we shall simplify matters by making the fairly natural assumption that all the capacities are integers and that $c(vv) = 0$ for each vertex v.

Example Let $N = (V, c)$ be the network with $V = \{u, v, w, x, y\}$ and with $c(uy) = 7$, $c(vw) = 1$, $c(vy) = 1$, $c(wu) = 3$, $c(wx) = 2$, $c(wy) = 2$, $c(xu) = 4$, $c(xv) = 3$, $c(xw) = 3$, and all other capacities zero.

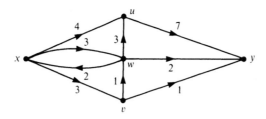

You can think of the edges as pipes where a current is allowed in the direction shown up to the given capacity. You can then construct various 'flows' through the network:

e.g.

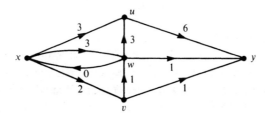

□

The 'flow of 8' from x to y illustrated in that example has several key properties: at x there is, in an obvious sense, an overall flow of 8 into the network and at y there is an overall flow of 8 out of the network, whereas at all other vertices there is no overall flow, the current 'in' equalling the current 'out'. Also the flow in each edge never exceeds the capacity of the edge. However, that flow is not the largest possible from x to y. Note that if you walk from x to y in the network along the route x, u, w, y then at each stage either you are walking the right way along an 'unsaturated' edge (i.e. one in which the flow is less than the capacity) or you are walking the wrong way along an edge which has an actual flow in it. Such a route is called 'a flow-augmenting path' because if you increase the flow by 1 in the edges you went along the right way and decrease the flow by 1 in the edges you went along the wrong way then you get a new flow in the network and the overall flow from x to y is increased by 1:

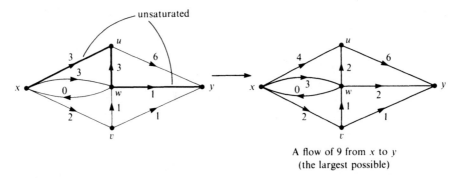

A flow of 9 from x to y
(the largest possible)

Returning to the original network, it is clearly possible to stop all flows from x to y by cutting through some of the edges. For example the simplest cuts to make would be of all the edges out of x, namely xv, xw and xu: the sum of their capacities is $3 + 3 + 4 = 10$. However, it is possible to stop all flow from x to y by cutting a set of edges with a lower total capacity than that: the set of edges xw, xu, vw and vy have total capacity $3 + 4 + 1 + 1 = 9$ and if they are cut they certainly stop all flow from x to y. Furthermore no set of cut edges of any lower total capacity will stop the flow from x to y.

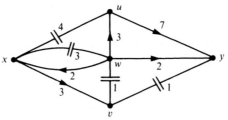

Formally, given a network $N = (V, c)$, *a flow in the edge vw is a number $f(vw)$* satisfying $0 \leqslant f(vw) \leqslant c(vw)$: this can be thought of as a flow of liquid in the pipe vw, out of v and into w, which must not exceed the capacity of that pipe. (Again we shall restrict attention to the cases when $f(vw)$ is an integer.) If overall the flows in the edges satisfy

Minimax theorems

(i) the flow of each edge into x is zero,
(ii) at each vertex v apart from x and y the total flows of the edges into v equals the total flows of the edges out of v, and
(iii) the flow of each edge out of y is zero,
then the function f (from $V \times V$ to the non-negative integers) is called *a flow from x to y* in the network: furthermore, x is called the *source* of the flow, and y is called the *sink*. Given such a flow, f, it is easy to see that the total of the flows of the edges out of x is the same as the total of the flows of the edges into y: this number is called the *value* of the flow. Between any two vertices of a network there is always the trivial example of a flow of value zero, where the flow in each edge is zero.

On the other hand a *cut* in $N = (V, c)$ separating x from y is a set of edges whose removal would leave a network where no non-trivial flow from x to y was possible. The *capacity* of the cut is the sum of the capacities of the edges in the cut.

In the above example the flows from x to y had largest value 9 which coincided with the smallest capacity of cuts, and of course that was not a coincidence: the general result, known as the 'max flow–min cut' theorem, was first proved by L.R. Ford and D.R. Fulkerson in 1956.

Theorem (Max flow–min cut) *Between any two vertices x and y in a network*

$$\begin{array}{c} \text{the minimum capacity of a} \\ \text{cut which separates } x \text{ from } y \end{array} = \begin{array}{c} \text{the maximum value of} \\ \text{a flow from } x \text{ to } y. \end{array}$$

Proof Let the network be $N = (V, c)$, let α be the minimum capacity of all possible cuts which separate x from y in N and let β be the maximum value of all possible flows from x to y. If a cut has capacity k then the removal of those edges stops all flow from x to y: hence any flow which is possible after the reinstatement of those edges must clearly have value k or less. Therefore the value of *any* flow is less than or equal to the capacity of *any* cut. In particular the value of the 'maximum flow' is less than or equal to the capacity of the 'minimum cut'; i.e. $\beta \leq \alpha$.

To show that $\alpha \leq \beta$ we shall show by induction on n that for $0 \leq n \leq \alpha$ there is a flow of value n. In particular that will show that there exists a flow of value α and that β (being the maximum flow) is greater than or equal to α.

Clearly there is a flow of value zero. Assume that $0 \leq n < \alpha$ and that a flow f of value n is given. We shall show that there exists a flow of value $n + 1$. Let W be the set of vertices w for which there exists a 'flow-augmenting path' from x to w; i.e. a sequence of vertices

$$x = x_0, x_1, x_2, \ldots, x_{m-1}, x_m = w$$

with the property that for each i with $0 \leq i < m$

either $\underbrace{f(x_i x_{i+1}) < c(x_i x_{i+1})}_{\substack{(x_i x_{i+1}) \text{ is in the network} \\ \text{in that direction and it} \\ \text{is 'unsaturated'}}}$ or $\underbrace{f(x_{i+1} x_i) > 0.}_{\substack{\text{(you use } x_i x_{i+1} \text{ in the wrong} \\ \text{direction and there is an} \\ \text{actual flow in the edge)}}}$

We claim that $y \in W$. For assume otherwise and consider the sets of vertices W (containing x) and $V \setminus W$ (containing y). Let C be the set of edges from W to $V \setminus W$ in that direction and let D be the set of edges from $V \setminus W$ to W in that direction. It is clear that the set of edges C form a cut separating x from y.

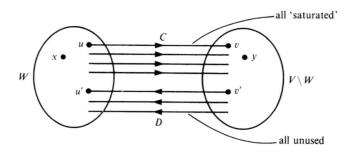

By the definition of W it follows that it is impossible to go from a vertex in W to a vertex in $V \setminus W$ along a flow-augmenting path. Hence every edge uv in C has $f(uv) = c(uv)$ and every edge $v'u'$ in D has $f(v'u') = 0$. The value of the flow (which is n and is less than α) clearly equals the total flow across from W to $V \setminus W$ which equals the total capacity of the edges in C. It therefore follows that C is a cut of capacity less than α, which contradicts the minimality of α. This contradiction shows that $y \in W$ as claimed.

Therefore, by the definition of W, there exists a flow-augmenting path from x to y; i.e. a sequence of vertices

$$x = x_0, x_1, x_2, \ldots, x_{m-1}, x_m = y$$

with the property that for each i with $0 \leq i < m$ either $f(x_i, x_{i+1}) < c(x_i, x_{i+1})$ or $f(x_{i+1}, x_i) > 0$. Now change the flow along the edges in the flow-augmenting path in the following ways:

$$\underbrace{\text{if } f(x_i x_{i+1}) < c(x_i x_{i+1}):}_{\substack{\text{for all edges } x_i x_{i+1} \\ \text{of this type increase} \\ \text{the flow by 1}}} \quad \text{or} \quad \underbrace{\text{if } f(x_{i+1} x_i) > 0:}_{\substack{\text{for all edges } x_{i+1} x_i \\ \text{of this type decrease} \\ \text{the flow by 1}}}$$

It is easy to check that this gives a new flow f' from x to y of value $n + 1$. We have therefore proved by induction that flows of all values from 0 to α exist. In particular, as commented earlier, the flow of value α shows that $\beta \geq \alpha$. Hence $\alpha = \beta$ and the max flow–min cut theorem is proved. □

That proof was algorithmic in the sense that, given any two vertices in a network, it is possible to use that method to find a maximum flow (and minimum cut).

Minimax theorems

Example *Find a maximum flow from x to y in the network illustrated.*

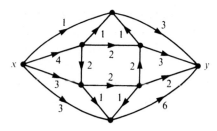

Solution We start with zero flow and in each case (until the last) show a flow-augmenting path which is then used to proceed to the next picture:

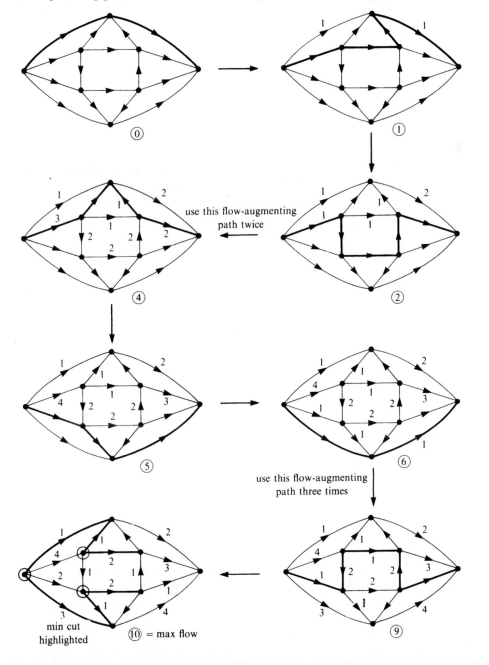

min cut highlighted ⑩ = max flow

There are no more flow-augmenting paths from x to y: the set W (as in the proof) is shown ringed in the final figure. So the maximum flow is 10 and a minimum cut of 10 is formed by the edges from W to $V \setminus W$ as shown in bold print. □

That brings to an end our discussion of minimax theorems but it is worth commenting that the ones which we have discussed (and some others which we have not) are closely related. For example we used Hall's theorem to deduce the König–Egerváry theorem which was a special case of Menger's theorem and, in some senses, the 'max flow–min cut' theorem is a generalised edge-form of Menger's theorem. Too much introversion would be inappropriate in an introductory course but we do see a couple of other easy connections in the exercises and, for a fuller story, the interested reader is referred to *The equivalence of some combinatorial matching theorems* by P.F. Reichmeider.

Exercises

1. Let $G = (V_1, E, V_2)$ be a bipartite graph. Use the König–Egerváry theorem to show that

   $$\begin{matrix}\text{the minimum number of vertices} \\ \text{which between them include at} \\ \text{least one endpoint of each edge}\end{matrix} = \begin{matrix}\text{the maximum number of} \\ \text{edges in a matching.}\end{matrix}$$ [H]

2. Let $G = (V_1, E, V_2)$ be a bipartite graph and for each $W \subseteq V_1$ let $j(W)$ be the set of vertices (in V_2) which are joined to at least one member of W. Show that the maximum number of edges in a matching in G is

   $$\min_{W \subseteq V_1} \{|V_1| - |W| + |j(W)|\}.$$ [H]

3. Deduce Hall's theorem from the König–Egerváry theorem. [H]

4. (The edge form of Menger's theorem.) Let $G = (V, E)$ be a graph and let x, y be vertices of G. A collection of paths from x to y is *edge-disjoint* if no edge is used in more than one of them, and a set of edges *separates* x from y if every path from x to y uses at least one of those edges. Use the 'max flow–min cut' theorem to show that the maximum number of edge-disjoint paths from x to y equals the minimum number of edges which separates x from y. [H]

10

Recurrence

In this chapter we return to the idea of counting various combinations and selections of items but here, instead of using direct arguments, we shall deduce our answers by building up to them from smaller situations.

Example *The famous Fibonacci sequence is $0, 1, 1, 2, 3, 5, 8, 13, 21, \ldots$, where, after the first two terms are given, subsequent terms are obtained by adding the two previous ones; i.e.*

$$F_0 = 0 \quad F_1 = 1 \quad \text{and} \quad F_n = F_{n-1} + F_{n-2} \quad (n > 1).$$

These relationships uniquely define the sequence and they enable you to calculate any particular term. Also, if you suspect that you have a formula for F_n, then if you simply check that it satisfies the above relationships it follows that it must be correct. For example it is left as an exercise to show that the formula

$$F_n = \frac{1}{\sqrt{5}} \left(\left(\frac{1 + \sqrt{5}}{2} \right)^n - \left(\frac{1 - \sqrt{5}}{2} \right)^n \right)$$

satisfies the relationships and hence gives a correct formula for F_n. But we do not yet know how those relationships can lead us to such a formula. □

Example *In chapter 1 we met Pascal's triangle:*

```
                1
              1   1
            1   2   1
          1   3   3   1
        1   4   6   4   1
      1   5  10  10   5   1
                 ⋮
```

Let us denote the entry in the first row by p_{00}, those in the next row by p_{10}, p_{01}, the next row by p_{20}, p_{11}, p_{02}, and so on, giving the general row as

$$p_{n0}, \quad p_{(n-1)1}, \quad p_{(n-2)2}, \quad \ldots, \quad p_{0n}.$$

Of course we now know what each entry is in terms of factorials, but imagine instead that you are just given the Pascal's triangle array and are asked to find a formula for p_{jk}. That is far from easy, but what is clear is the way in which the pattern is defined. After the two outer lines of 1s are in place each term is obtained by adding together the two terms immediately above it; i.e.

$$p_{n0} = 1 \qquad p_{0n} = 1 \qquad \text{and} \qquad p_{jk} = p_{(j-1)k} + p_{j(k-1)} \qquad (j, k > 0).$$

Those relationships certainly uniquely determine each entry in the triangle and they enable you to work out any specific one. For example

$$p_{22} = p_{12} + p_{21} = (p_{02} + p_{11}) + (p_{11} + p_{20}) = 2 + 2p_{11} = 2 + 2(p_{01} + p_{10}) = 6.$$
$$\qquad\qquad\qquad\;\; \| \qquad\qquad\qquad \| \qquad\qquad\qquad\qquad\qquad\;\; \| \quad \|$$
$$\qquad\qquad\qquad\;\; 1 \qquad\qquad\qquad 1 \qquad\qquad\qquad\qquad\qquad\;\; 1 \quad 1$$

Also, as in the previous example, if you suspect that you have a formula for p_{jk} then you must simply check that it satisfies the above relationships. However without some new techniques the relationships are no help in actually finding a general formula for p_{jk}. □

In both those examples the numbers are defined by a *recurrence relation* where, after some initial terms are given, a formula tells you how subsequent terms are generated. In some cases it is possible to 'solve' the recurrence relation; i.e. to use it to obtain an explicit formula for the general term. This chapter consists of a selection of examples of recurrence relations illustrating various techniques for solving them. The next two examples are reasonably straightforward because the recurrence relation in each case simply uses one previous term (e.g. it expresses m_n using only m_{n-1}).

Example (The Tower of Hanoi) *In this famous puzzle (sometimes known as The Tower of Brahma) you are given a stand with three vertical rods. On the first is a pile of n rings with decreasing radii, as illustrated. Your task is to move the pile to the third rod: you can only move one ring at a time, from one rod to another, and in each pile no ring may ever be on top of a smaller one. What is the minimum number of moves required?*

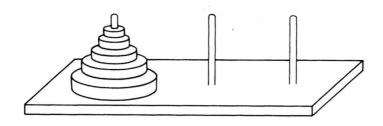

Recurrence

Solution Let m_n be the minimum number of moves required in the case of n rings. Clearly $m_1 = 1$. Now imagine, for example, that you have learned how to move four rings in the most economical way, taking m_4 moves. Then you try the same problem with five rings. You can move the top four rings to the middle rod (taking m_4 moves), the largest ring to the third rod (one more move), and the four rings from the middle rod to the third rod (taking a further m_4 moves).

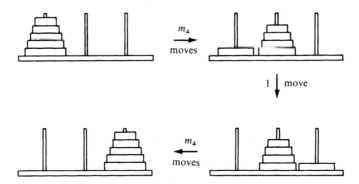

Also, to move the five rings you are forced to move the top four off first and so the above method, taking $2m_4 + 1$ moves, is the most economical. Hence $m_5 = 2m_4 + 1$. Of course the argument extends to n rings to give the recurrence relation

$$m_1 = 1 \quad \text{and} \quad m_n = 2m_{n-1} + 1 \quad (n > 1).$$

Hence $m_2 = 2m_1 + 1 = 3$, $m_3 = 2m_2 + 1 = 7, \ldots$ and this rule generates the sequence $1, 3, 7, 15, 31, 63, \ldots$. You can probably spot the pattern, namely $m_n = 2^n - 1$. The recurrence relation uniquely determines the sequence and so to confirm that this formula is correct we must simply note that it does satisfy the relation:

$$m_1 = 2^1 - 1 = 1 \quad \text{and} \quad m_n = 2^n - 1 = 2(2^{n-1} - 1) + 1 = 2m_{n-1} + 1.$$

Hence the number of moves needed for n rings in the Tower of Hanoi puzzle is indeed $2^n - 1$. □

Example *In chapter 2 we used the multinomial coefficients to prove Cayley's theorem that the number of trees on vertex-set $\{1, 2, \ldots, n\}$ is n^{n-2}. An alternative method (a version of which can be found in R.J. Wilson's* Introduction to graph theory*) leads to a relationship concerning the number of such trees in which the degree of the first vertex is specified. In particular, if t_k denotes the number of trees on vertex-set $\{1, 2, \ldots, n\}$ with vertex 1 joined to all except k of the other vertices (i.e. with the degree of vertex 1 equal to $n - k - 1$), then it can be shown that*

$$t_k = \frac{(n-k-1)(n-1)}{k} t_{k-1} \quad (0 < k < n - 1).$$

Find an explicit formula for t_k and show that

$$t_0 + t_1 + t_2 + \cdots + t_{n-2} = n^{n-2}.$$

Solution Note firstly that t_0 is the number of trees on vertex-set $\{1, 2, \ldots, n\}$ with vertex 1 joined to all the other vertices. There is clearly only one such tree, as illustrated, and hence $t_0 = 1$. That starting-point together with the formula for t_k in terms of t_{k-1} gives us a recurrence relation which uniquely determines the values of $t_0, t_1, t_2, \ldots$ and t_{n-2}. By applying the relation in the k-case and then in the $(k-1)$-case etc., we can deduce that

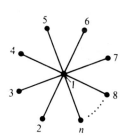

$$t_k = \frac{(n-k-1)(n-1)}{k} t_{k-1} = \frac{(n-k-1)(n-1)}{k} \frac{(n-(k-1)-1)(n-1)}{k-1} t_{k-2}$$

$$= \frac{(n-k-1)(n-k)(n-1)^2}{k(k-1)} t_{k-2}$$

$$= \frac{(n-k-1)(n-k)(n-k+1)(n-1)^3}{k(k-1)(k-2)} t_{k-3}$$

$$\vdots$$

$$= \frac{(n-k-1)(n-k)\cdots(n-2)(n-1)^k}{k(k-1)\cdots 1} t_0$$

$$= \binom{n-2}{k}(n-1)^k.$$

Hence

$$t_0 + t_1 + t_2 + \cdots + t_{n-2}$$

$$= \underbrace{\binom{n-2}{0} + \binom{n-2}{1}(n-1) + \binom{n-2}{2}(n-1)^2 + \cdots + \binom{n-2}{n-2}(n-1)^{n-2}}_{\text{the binomial expansion of } (1+(n-1))^{n-2}}$$

$$= n^{n-2}. \qquad \square$$

In both those examples the recurrence relation simply involved one previous term but we now return to examples where, as in the case of the Fibonacci numbers, the recurrence relation involves two previous terms.

Theorem *Let a_0 and a_1 be given and let $a_2, a_3, \ldots$ be defined by the recurrence relation*

$$a_n = Aa_{n-1} + Ba_{n-2} \qquad (n > 1),$$

where A and B are constants. Then let α, β be the roots of the quadratic equation $x^2 = Ax + B$ (known as the auxiliary equation). It then follows that

(i) if $\alpha \neq \beta$ there exist constants c, d with
$$a_n = c\alpha^n + d\beta^n \quad \text{for each } n \geq 0,$$

(ii) if $\alpha = \beta$ there exist constants c, d with
$$a_n = (c + dn)\alpha^n \quad \text{for each } n \geq 0.$$

Proof We prove (i) (the proof of (ii) is very similar and is left as an exercise). It is easy to check that, whatever the value of the constants c and d, the given formula satisfies $a_n = Aa_{n-1} + Ba_{n-2}$. For
$$Aa_{n-1} + Ba_{n-2} = A(c\alpha^{n-1} + d\beta^{n-1}) + B(c\alpha^{n-2} + d\beta^{n-2})$$
$$= c\alpha^{n-2}\underbrace{(A\alpha + B)}_{=\alpha^2} + d\beta^{n-2}\underbrace{(A\beta + B)}_{=\beta^2}$$
$$= c\alpha^n + d\beta^n$$
$$= a_n.$$

That general step depends upon the fact that α and β are the roots of $x^2 = Ax + B$ but it does not depend upon the values of c and d: they must now be chosen to make the formula work in the cases $n = 0$ and $n = 1$; i.e. to satisfy
$$c + d(=c\alpha^0 + d\beta^0) = a_0 \quad \text{and} \quad c\alpha + d\beta = a_1.$$

Since $\alpha \neq \beta$ these equations determine unique values of c and d and, for those particular values, it follows that the given formula holds for all $n \geq 0$. □

Example *Find a formula for the Fibonacci number F_n.*

Solution We know that $F_0 = 0$, $F_1 = 1$, and $F_n = F_{n-1} + F_{n-2}$ for $n > 1$. So this recurrence relation has auxiliary equation $x^2 = x + 1$, which has roots $\frac{1}{2}(1 + \sqrt{5})$ and $\frac{1}{2}(1 - \sqrt{5})$. Therefore, by the theorem,
$$F_n = c\left(\frac{1 + \sqrt{5}}{2}\right)^n + d\left(\frac{1 - \sqrt{5}}{2}\right)^n$$

for some constants c and d. To find the value of the constants use the fact that the formula works for $n = 0$ and $n = 1$ to give
$$c + d = F_0 = 0 \quad \text{and} \quad c\frac{1 + \sqrt{5}}{2} + d\frac{1 - \sqrt{5}}{2} = F_1 = 1:$$

these have solution $c = 1/\sqrt{5}$ and $d = -1/\sqrt{5}$. Hence
$$F_n = \frac{1}{\sqrt{5}}\left(\left(\frac{1 + \sqrt{5}}{2}\right)^n - \left(\frac{1 - \sqrt{5}}{2}\right)^n\right)$$

(which is the formula we had before). □

That theorem (and simple generalisations of it) enable us to solve recurrence relations in which each term is expressed as a sum of *constant* multiples of a *fixed number* of previous terms, whereas most of the interesting recurrence relations encountered in combinatorics are less straightforward than that. In the next example we introduce another method of solution of a simple recurrence relation which can in fact cope with a wide variety of problems: this particular example finds a different formulation of the Fibonacci numbers.

Example *Let $F_0, F_1, F_2, \ldots$ be the Fibonacci numbers and let f be the function defined by*

$$f(x) = F_0 + F_1 x + F_2 x^2 + \cdots.$$

Show that

$$f(x) = \frac{x}{1 - x(1 + x)}$$

and hence express F_n as a sum of bionomial coefficients.

Solution As mathematicians you might be thinking about convergence and wondering for which x the expression $f(x)$ is well-defined, but actually that need not concern us. The function f is called the 'generating function' of the sequence $F_0, F_1, F_2, \ldots$: it is a 'formal' algebraic expression in the sense that it behaves in a natural algebraic way but the x never takes a numerical value, the powers of it merely acting as 'place-markers' in the power series. Bearing that in mind, we see that

$$f(x) = F_0 + F_1 x + F_2 x^2 + F_3 x^3 + \cdots,$$
$$xf(x) = F_0 x + F_1 x^2 + F_2 x^3 + \cdots,$$
$$x^2 f(x) = F_0 x^2 + F_1 x^3 + \cdots.$$

But we know that $F_2 = F_1 + F_0$, $F_3 = F_2 + F_1$, etc., and so subtracting the second and third lines from the first gives us

$$f(x)(1 - x(1 + x)) = f(x) - xf(x) - x^2 f(x) = F_0 + (F_1 - F_0)x = x$$
$$ \| \| \|$$
$$ 0 1 0$$

and hence that

$$f(x) = \frac{x}{1 - x(1 + x)}$$

as required. The series expansion of $1/(1 - y)$ is the geometric series

$$1 + y + y^2 + y^3 + \cdots$$

Recurrence

and so putting $y = x(1 + x)$ gives us

$$f(x) = x[1 + x(1 + x) + x^2(1 + x)^2 + x^3(1 + x)^3 + \cdots].$$

Now F_n is simply the coefficient of x^n in $f(x)$ and so to retrieve F_n we must count the number of x^ns in our new expansion of $f(x)$. Therefore we need to count the x^{n-1}s in each $x^k(1 + x)^k$: that equals the number of x^{n-k-1}s in $(1 + x)^k$ and that is the binomial coefficient $\binom{k}{n - k - 1}$. Hence the number of x^n in the expansion is as shown below:

$f(x) =$

$x[1 + \cdots + x^{n-3}(1 + x)^{n-3} + x^{n-2}(1 + x)^{n-2} + x^{n-1}(1 + x)^{n-1} + \cdots \qquad].$

$\qquad \cdots \quad \binom{n-3}{2} \qquad \binom{n-2}{1} \qquad \binom{n-1}{0} \qquad$ zero hereafter

It follows that

$$F_n = \binom{n-1}{0} + \binom{n-2}{1} + \binom{n-3}{2} + \cdots$$

(where these binomial coefficients are all eventually zero). □

Given a sequence $a_0, a_1, a_2, \ldots$ its *generating function* is given by

$$a(x) = a_0 + a_1 x + a_2 x^2 + \cdots.$$

In the previous example we commented about the purely algebraic nature of the generating function and we began to see some of its uses: the next few examples demonstrate its versatility.

Example Let N be a fixed positive integer and for $0 \leq n \leq N$ let

$$a_n = \binom{n}{n} + \binom{n+1}{n} + \binom{n+2}{n} + \cdots + \binom{N}{n}.$$

Use the method of generating functions to express a_n as a single binomial coefficient.

Solution We established a binomial identity equivalent to this one in an example on page 4 but here we use it to illustrate the use of generating functions. We wish to find $a_0, a_1, \ldots$ and a_N and so we define the generating function of this sequence and see what it looks like when written out in full:

$$a(x) = a_0 + a_1 x + a_2 x + \cdots + a_N x^N$$

$$= \binom{0}{0} + \binom{1}{0} + \binom{2}{0} + \binom{3}{0} + \binom{4}{0} + \cdots + \binom{N}{0}$$

$$+ \binom{1}{1}x + \binom{2}{1}x + \binom{3}{1}x + \binom{4}{1}x + \cdots + \binom{N}{1}x$$

$$+ \binom{2}{2}x^2 + \binom{3}{2}x^2 + \binom{4}{2}x^2 + \cdots + \binom{N}{2}x^2$$

$$+ \binom{3}{3}x^3 + \binom{4}{3}x^3 + \cdots + \binom{N}{3}x^3$$

$$\ddots \qquad \vdots$$

$$+ \binom{N}{N}x^N.$$

Then each column is a simple binomial expansion and so we can add up this grand total column by column to give

$$a(x) = 1 + (1 + x) + (1 + x)^2 + (1 + x)^3 + \cdots + (1 + x)^N.$$

This is a geometric progression with sum

$$a(x) = \frac{(1+x)^{N+1} - 1}{(1+x) - 1} = \frac{(1+x)^{N+1} - 1}{x}.$$

Now a_n is the coefficient of x^n in this expression, which equals the coefficient of x^{n+1} in $(1 + x)^{N+1}$, namely $\binom{N+1}{n+1}$. □

Example (i) *Show that the set* $\{1, 2, \ldots, n\}$ *can be partitioned into two non-empty sets in precisely* $2^{n-1} - 1$ *ways.*

(ii) *Let* s_n *denote the number of ways in which the set* $\{1, 2, \ldots, n\}$ *can be partitioned into three non-empty sets. For example* $s_4 = 6$, *the six partitions being*

$$\{1\} \ \{2\} \ \{3, 4\}, \qquad \{1\} \ \{3\} \ \{2, 4\}, \qquad \{1\} \ \{4\} \ \{2, 3\},$$
$$\{2\} \ \{3\} \ \{1, 4\}, \qquad \{2\} \ \{4\} \ \{1, 3\}, \qquad \{3\} \ \{4\} \ \{1, 2\}.$$

Show that s_n *is determined by the recurrence relation*

$$s_0 = 0 \qquad s_1 = 0 \quad \text{and} \quad s_n = 3s_{n-1} + 2^{n-2} - 1 \qquad (n > 1).$$

Use a generating function to find an explicit formula for s_n.

Solution (i) The order of the two sets is irrelevant and so let n be in the first of the two sets. Then partitioning $\{1, 2, \ldots, n\}$ into two non-empty sets is clearly equivalent to choosing the second set as any non-empty subset of $\{1, 2, \ldots, n-1\}$: there are $2^{n-1} - 1$ such sets.

(ii) Now consider the s_n partitions of $\{1, 2, \ldots, n\}$ into three sets: either the set $\{n\}$ will be one of the three sets or else it will not! In the former case the set $\{1, 2, \ldots, n-1\}$ can be partitioned into the two remaining sets in $2^{n-2} - 1$ ways (by (i)): in the latter case the set $\{1, 2, \ldots, n-1\}$ must be partitioned into three sets and n placed in any one of them, giving $3s_{n-1}$ partitions of that type. It then follows easily that

$$s_0 = 0 \qquad s_1 = 0 \qquad \text{and} \qquad s_n = 3s_{n-1} + 2^{n-2} - 1 \qquad (n > 1).$$

Let s be the generating function of the sequence $s_0, s_1, s_2, \ldots$; i.e.

$$s(x) = s_0 + s_1 x + s_2 x^2 + s_3 x^3 + s_4 x^4 + \cdots$$

Then

$$3xs(x) = \qquad 3s_0 x + 3s_1 x^2 + 3s_2 x^3 + 3s_3 x^4 + \cdots$$

But, by the above recurrence relation, $s_0 = s_1 = 0$, $s_2 - 3s_1 = 2^0 - 1$, $s_3 - 3s_2 = 2^1 - 1$, etc. Hence subtracting $3xs(x)$ from $s(x)$ gives

$$(1 - 3x)s(x) = (2^0 - 1)x^2 + (2^1 - 1)x^3 + (2^2 - 1)x^4 + \cdots$$

$$= x^2[(1 + (2x) + (2x)^2 + (2x)^3 + \cdots) - (1 + x + x^2 + x^3 + \cdots)]$$

$$= x^2 \left(\frac{1}{1 - 2x} - \frac{1}{1 - x} \right).$$

Therefore, after a little manipulation with partial fractions, we see that

$$s(x) = x^2 \left(\frac{1}{(1 - 2x)(1 - 3x)} - \frac{1}{(1 - x)(1 - 3x)} \right)$$

$$= x^2 \left(\frac{3}{2(1 - 3x)} + \frac{1}{2(1 - x)} - \frac{2}{1 - 2x} \right).$$

We can now deduce that s_n, being the coefficient of x^n in $s(x)$, is given by

$$s_n = \tfrac{3}{2} 3^{n-2} + \tfrac{1}{2} - 2 \cdot 2^{n-2} = \tfrac{1}{2}(3^{n-1} + 1 - 2^n).$$

(The number of different ways of partitioning $\{1, 2, \ldots, n\}$ into k non-empty sets is denoted by $S(n, k)$ and is called a *Stirling number of the second kind* after the eighteenth century mathematician James Stirling. We shall derive a recurrence relation for these numbers – and for those of the 'first kind' – in the exercises.) □

Example *Given that the numbers p_{jk} in Pascal's triangle (as in the earlier example) satisfy the recurrence relation*

$$p_{n0} = 1 \qquad p_{0n} = 1 \qquad \text{and} \qquad p_{jk} = p_{(j-1)k} + p_{j(k-1)} \qquad (j, k > 0)$$

find an explicit formula for p_{jk}.

Solution This example is rather artificial because we already know a formula for p_{jk} (namely $(j+k)!/j!k!$) and we could simply check that this formula does indeed satisfy the given recurrence relation. But we shall imagine that we do not yet know a formula for p_{jk} and we shall use the method of generating functions to find one.

There are several ways of dealing with a 'double sequence' like the p_{jk}s where there are two variables: we shall define a generating function for each row of Pascal's triangle. So for $n \geq 0$ let $f_n(x)$ be given by

$$f_n(x) = p_{n0} + p_{(n-1)1}x + p_{(n-2)2}x^2 + \cdots + p_{1(n-1)}x^{n-1} + p_{0n}x^n.$$

Then

$$f_1(x) = p_{10} + p_{01}x = 1 + x$$

and we can use the given recurrence relation to see that for $n > 1$

$$\begin{aligned} f_n(x) &= \quad p_{n0} \quad + \quad p_{(n-1)1}x \quad + \quad p_{(n-2)2}x^2 \quad + \cdots + \quad p_{0n}x^n \\ &= \quad 1 \quad + (p_{(n-2)1} + p_{(n-1)0})x + (p_{(n-3)2} + p_{(n-2)1})x^2 + \cdots + \quad 1x^n \\ & \quad \| \qquad\qquad\qquad\qquad\qquad\qquad\qquad\qquad\qquad\qquad\qquad\qquad\qquad\qquad \| \\ & \quad p_{(n-1)0} \qquad\qquad\qquad\qquad\qquad\qquad\qquad\qquad\qquad\qquad\qquad\qquad\qquad p_{0(n-1)} \end{aligned}$$

$$= (1+x)(p_{(n-1)0} + p_{(n-2)1}x + p_{(n-3)2}x^2 + \cdots + p_{0(n-1)}x^{n-1})$$

$$= (1+x)f_{n-1}(x).$$

Hence the generating functions f_n themselves satisfy a simple recurrence relation, namely

$$f_1(x) = 1 + x \quad \text{and} \quad f_n(x) = (1+x)f_{n-1}(x) \quad (n > 1).$$

Therefore

$$f_n(x) = (1+x)f_{n-1}(x) = (1+x)^2 f_{n-2}(x) = \cdots = (1+x)^{n-1}f_1(x) = (1+x)^n$$

and $f_n(x)$ is simply $(1+x)^n$. So if $j + k = n$ then the entry p_{jk} in Pascal's triangle, being the coefficient of x^k in $f_n(x)$, is merely the coefficient of x^k in $(1+x)^n$; i.e. $\binom{n}{k}$.

That is, after all, no surprise but it does show the generating functions at work in a very neat way. □

Example At the time of writing, the British coins readily available are the 1p, 2p, 5p, 10p, 20p, 50p and £1. Show that the number of different ways of making a total of n pence with limitless supplies of these coins equals the coefficient of x^n in

$$((1-x)(1-x^2)(1-x^5)(1-x^{10})(1-x^{20})(1-x^{50})(1-x^{100}))^{-1}.$$

Recurrence

Solution The given function can be written as

$$(1 + x + x^2 + x^3 + \cdots)(1 + x^2 + x^4 + \cdots)(1 + x^5 + x^{10} + \cdots)$$
$$\times (1 + x^{10} + x^{20} + \cdots)(1 + x^{20} + x^{40} + \cdots)(1 + x^{50} + \cdots)(1 + x^{100} + \cdots).$$

How, for example, can an x^{21} occur in this expansion? A few examples are

$$x^{21} \cdot 1 \cdot 1 \cdot 1 \cdot 1 \cdot 1 \cdot 1 \quad \text{or} \quad x^{19} \cdot x^2 \cdot 1 \cdot 1 \cdot 1 \cdot 1 \cdot 1 \quad \text{or} \quad x^{11} \cdot 1 \cdot x^{10} \cdot 1 \cdot 1 \cdot 1 \cdot 1$$
$$\text{or} \quad x^{11} \cdot 1 \cdot 1 \cdot x^{10} \cdot 1 \cdot 1 \cdot 1 \quad \text{or} \quad \cdots.$$

Each product consists of seven components. If the first component is x^a then you can think of it as a pence in 1p coins: if the second component is x^b then you can think of it as b pence in 2p coins, etc. So the above products correspond to

$$21 \text{ @ } 1p \quad \text{or} \quad 19 \text{ @ } 1p, 1 \text{ @ } 2p \quad \text{or} \quad 11 \text{ @ } 1p, 2 \text{ @ } 5p$$
$$\text{or} \quad 11 \text{ @ } 1p, 1 \text{ @ } 10p \quad \text{or} \quad \cdots$$

A little thought will show you that there is a one-to-one correspondence between the ways of getting an x^{21} in the expansion and the ways of making 21p in coins; i.e. the coefficient of x^{21} equals the number of ways of making 21p in coins: of course a similar argument works for n pence. □

Example Let d_n denote the number of ways of obtaining a total of n when successively throwing a die. For example $d_4 = 8$, the possible throws being

$$1 + 1 + 1 + 1, \quad 1 + 1 + 2, \quad 1 + 2 + 1, \quad 2 + 1 + 1,$$
$$1 + 3, \quad 2 + 2, \quad 3 + 1, \quad 4.$$

Show that d_n is the coefficient of x^n in $(1 - x - x^2 - x^3 - x^4 - x^5 - x^6)^{-1}$.

Solution It is worth noting an important difference between this example and the previous one, namely that order matters. Here the total of 4 obtained as $1 + 2 + 1$ is regarded as different from $1 + 1 + 2$, whereas when making a total of 4p with two 1p and one 2p the order of the coins was irrelevant.

One way of tackling this exercise is to find a recurrence relation for the sequence $d_0, d_1, d_2, \ldots$ and then to use our usual method for deducing the generating function $d(x)$. For example it is not hard to see (by considering the last throw) that

$$d_n = d_{n-1} + d_{n-2} + d_{n-3} + d_{n-4} + d_{n-5} + d_{n-6} \quad (n > 6)$$

and to use it, together with some initial values, to deduce the required form of $d(x)$.

Alternatively we can begin to be a little more inspired with our use of generating functions. Clearly the number of ways of throwing a total of n with *one* throw of the die equals the coefficient of x^n in

$$x + x^2 + x^3 + x^4 + x^5 + x^6$$

(because you can throw 1–6 in just one way and all other totals in zero ways). A moment's thought will show you that the number of ways of throwing a total of n with *two* throws of the die equals the coefficient of x^n in

$$(x + x^2 + x^3 + x^4 + x^5 + x^6)(x + x^2 + x^3 + x^4 + x^5 + x^6)$$

(for example you can get a total of 5 as $1 + 4$ or $2 + 3$ or $3 + 2$ or $4 + 1$, and you can get x^5s as $x \cdot x^4$ or $x^2 \cdot x^3$ or $x^3 \cdot x^2$ or $x^4 \cdot x$). It is not hard to extend this to the fact that the number of ways of throwing a total of n with k throws of the die equals the coefficient of x^n in

$$(x + x^2 + x^3 + x^4 + x^5 + x^6)^k \qquad \text{(any } k \geqslant 0\text{)}.$$

Hence the number of ways of throwing a total of n when *any* number of throws is allowed is

$$\sum_{k=0}^{\infty} \text{coefficient of } x^n \text{ in } (x + x^2 + x^3 + x^4 + x^5 + x^6)^k$$

$$= \text{coefficient of } x^n \text{ in } \sum_{k=0}^{\infty} (x + x^2 + x^3 + x^4 + x^5 + x^6)^k$$

$$= \text{coefficient of } x^n \text{ in } (1 - (x + x^2 + x^3 + x^4 + x^5 + x^6))^{-1}$$

as required. □

We have already seen two examples of 'partitioning', the idea of partitioning a set into three subsets and the idea of partitioning n pence into subtotals made up of coins of various denominations. In number theory a *partition* of the positive integer n is a way of expressing n as a sum of positive integers. For example there are seven different partitions of 5, namely $5, 4 + 1, 3 + 2, 3 + 1 + 1, 2 + 2 + 1, 2 + 1 + 1 + 1, 1 + 1 + 1 + 1 + 1$. Note that the order of the sum is irrelevant, so that $3 + 2$ is regarded as the same partition of 5 as $2 + 3$. It is a difficult problem to find a formula for the number of partitions of n. Generating functions enable us to express the answer as a coefficient of an expansion and although it does not give us a very tangible answer it does enable us to learn some surprising facts about partitions.

Example Let p_n denote the number of different partitions of n, let d_n be the number of those partitions where the breakdown is into distinct integers, and let o_n be the number of partitions of n where only odd integers are used. So for example $p_5 = 7$ (as above), $d_5 = 3$ (only the partitions $5, 4 + 1$ and $3 + 2$ being allowed) and $o_5 = 3$ (only the partitions $5, 3 + 1 + 1$ and $1 + 1 + 1 + 1 + 1$ being allowed). Find generating functions for the p_ns, the d_ns and the o_ns, and deduce that $d_n = o_n$ for each n.

Solution We are familiar with the idea of counting partitions: in the coin example we had to find partitions of n in which only the numbers 1, 2, 5, 10, 20, 50 and 100 could be used. The same principle leads easily to a generating function for the p_ns, namely

$$p(x) = (1 + x + x^2 + x^3 + \cdots)(1 + x^2 + x^4 + x^6 + \cdots)(1 + x^3 + x^6 + x^9 + \cdots)\cdots.$$

$\underbrace{}_{\text{(counting 1s)}}$ $\underbrace{}_{\text{(counting 2s)}}$ $\underbrace{}_{\text{(counting 3s)}}$

If we are not allowed to use any integer more than once then this reduces to the generating function for the d_ns, namely

$$d(x) = (1 + x)(1 + x^2)(1 + x^3)\cdots;$$

and if we are only allowed to use odd integers we get the generating function for the o_ns, namely

$$o(x) = (1 + x + x^2 + x^3 + \cdots)(1 + x^3 + x^6 + x^8 + \cdots)\cdots.$$

Hence the three required generating functions are

$$p(x) = ((1 - x)(1 - x^2)(1 - x^3)(1 - x^4)(1 - x^5)\cdots)^{-1},$$
$$d(x) = (1 + x)(1 + x^2)(1 + x^3)(1 + x^4)(1 + x^5)\cdots,$$
$$o(x) = ((1 - x)(1 - x^3)(1 - x^5)(1 - x^7)(1 - x^9)\cdots)^{-1}.$$

Now to show that $d_n = o_n$ for each n we shall simply show that the two relevant generating functions (although they look rather different) are actually the same:

$$d(x) = (1 + x)(1 + x^2)(1 + x^3)(1 + x^4)(1 + x^5)\cdots$$

$$= \frac{1 - x^2}{1 - x} \cdot \frac{1 - x^4}{1 - x^2} \cdot \frac{1 - x^6}{1 - x^3} \cdot \frac{1 - x^8}{1 - x^4} \cdot \frac{1 - x^{10}}{1 - x^5} \cdots$$

$$= \frac{1}{1 - x} \cdot \frac{1}{1 - x^3} \cdot \frac{1}{1 - x^5} \cdots$$

$$= o(x).$$

Therefore the number of partitions of any n into distinct integers is indeed equal to the number of partitions of n into odd integers. □

That ends our first excursion into recursion and we shall explore these ideas further in the exercises. We shall also meet some more unusual recurrence relations in subsequent chapters.

Exercises

1. (i) For each positive integer n let g_n be the number of subsets of $\{1, 2, \ldots, n\}$ in which no two members differ by 1. So, for example, $g_3 = 5$, the five sets being $\emptyset$, $\{1\}$, $\{2\}$, $\{3\}$ and $\{1, 3\}$. Find a recurrence relation for $g_1, g_2, g_3, \ldots$ and hence express g_n as a Fibonacci number. [A]

(ii) For each positive integer n let j_n denote the number of ways of arranging n identical dominoes, each 2 cm × 1 cm, in a tray 2 cm × n cm with no overlapping. For example, $j_3 = 3$, the three arrangements being as shown:

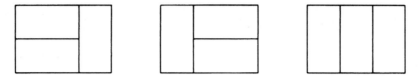

Find a recurrence relation for $j_1, j_2, j_3, \ldots$ and hence express j_n as a Fibonacci number. [A]

2. For each non-negative integer n let c_n denote the number of lists of n numbers each chosen from $\{0, 1, 2\}$ but with no two consecutive 1s and no two consecutive 2s. So, for example, $c_3 = 17$, the seventeen lists being

   000,   001,   002,   010,   012,   020,   021,   100,

   101,   102,   120,   121,   200,   201,   202,   210,   212.

   (i) Show that the c_ns satisfy the recurrence relation

   $$c_0 = 1, \quad c_1 = 3 \quad \text{and} \quad c_n = 2c_{n-1} + c_{n-2} \quad (n > 1). \quad [H]$$

   (ii) Use the theorem to solve the recurrence relation. Hence express c_n in terms of binomial coefficients and integer powers of 2. [A]

   (iii) Use the method of generating functions to find an alternative formula for c_n in terms of binomial coefficients and integer powers of 2. [A]

3. In the famous *Chinese ring puzzle*, rings and strings are threaded onto a bar so that although the first ring (i.e. the right-hand one in the figure below) can be removed freely, any other ring can be removed only when the one before it (i.e. the next to the right) is on the bar but all others before that are off the bar. So, for example, the fifth ring from the right can be removed only when the first, second and third ring are off the bar and the fourth is on it.

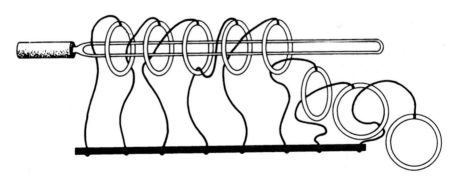

Let r_n be the minimum number of moves (i.e. moving one ring on or off the bar) required to remove n rings from the bar. Show that the r_ns satisfy the recurrence

relation
$$r_0 = 0, \quad r_1 = 1 \quad \text{and} \quad r_n = r_{n-1} + 2r_{n-2} + 1 \quad (n > 1).$$
Hence use a generating function to find a specific formula for r_n. [A]

4. For each positive integer k and non-negative integer n let $b_{nk} = \binom{n+k-1}{n}$ and let $b_k(x)$ be the generating function of the sequence $b_{0k}, b_{1k}, b_{2k}, \ldots$. Show that
$$b_1(x) = (1-x)^{-1} \quad \text{and} \quad b_k(x) = \frac{b_{k-1}(x)}{1-x} \quad (k > 1)$$
and deduce that b_{nk} is the coefficient of x^n in $(1-x)^{-k}$.

5. Let k and n be non-negative integers. Show that the number of different solutions of the equation
$$y_1 + y_2 + \cdots + y_k = n$$
in which each y_i is a non-negative integer equals the coefficient of x^n in $(1-x)^{-k}$, thus confirming the result met in chapter 1. [H]

6. (i) Let k and n be positive integers. Show that
$$\sum_{\substack{y_1, \ldots, y_k \text{ are} \\ \text{positive integers} \\ \text{with sum } n}} y_1 y_2 \cdots y_k$$
equals the coefficient of x^n in $(x + 2x^2 + 3x^3 + \cdots)^k$. Show that this last expression is $x^k(1-x)^{-2k}$ and hence express the above sum as a binomial coefficient. [A]

(ii) n children stand in a line (in a fixed order). Their teacher divides them into k groups, the first group being the first few children in the line, the second group being the next few, etc., and with each group consisting of at least one child. She then appoints a leader from each group. Use (i) to show that these groups and leaders can be chosen in $\binom{n+k-1}{2k-1}$ ways. Then find a direct selection argument to justify this answer. [H]

7. (*Inversion* – illustrated for even m.) Let M and N be the matrices shown below:

$$M = \begin{pmatrix} \binom{1}{1} & 0 & 0 & \cdots & 0 \\ \binom{2}{1} & \binom{2}{2} & 0 & \cdots & 0 \\ \binom{3}{1} & \binom{3}{2} & \binom{3}{3} & \cdots & 0 \\ \vdots & \vdots & \vdots & & \vdots \\ \binom{m}{1} & \binom{m}{2} & \binom{m}{3} & \cdots & \binom{m}{m} \end{pmatrix},$$

$$N = \begin{pmatrix} \binom{1}{1} & 0 & 0 & \cdots & 0 \\ -\binom{2}{1} & \binom{2}{2} & 0 & \cdots & 0 \\ \binom{3}{1} & -\binom{3}{2} & \binom{3}{3} & \cdots & 0 \\ \vdots & \vdots & \vdots & & \vdots \\ -\binom{m}{1} & \binom{m}{2} & -\binom{m}{3} & \cdots & \binom{m}{m} \end{pmatrix}.$$

Show that for $i \geqslant j$ the (i,j)th entry of the product MN is equal to the coefficient of x^{i-j} in $(1+x)^i (1+x)^{-(j+1)}$. Deduce that $M^{-1} = N$. [H]

Use that result to show that if

$$\binom{1}{1} a_1 = b_1,$$

$$\binom{2}{1} a_1 + \binom{2}{2} a_2 = b_2,$$

$$\binom{3}{1} a_1 + \binom{3}{2} a_2 + \binom{3}{3} a_3 = b_3,$$

$$\vdots \qquad \vdots \qquad \vdots \qquad \qquad \vdots$$

$$\binom{m}{1} a_1 + \binom{m}{2} a_2 + \binom{m}{3} a_3 + \cdots + \binom{m}{m} a_m = b_m,$$

then

$$\binom{1}{1} b_1 = a_1,$$

$$-\binom{2}{1} b_1 + \binom{2}{2} b_2 = a_2,$$

$$\binom{3}{1} b_1 - \binom{3}{2} b_2 + \binom{3}{3} b_3 = a_3,$$

$$\vdots \qquad \vdots \qquad \vdots \qquad \qquad \vdots$$

$$-\binom{m}{1} b_1 + \binom{m}{2} b_2 - \binom{m}{3} b_3 + \cdots + \binom{m}{m} b_m = a_m.$$

8. Let m be a fixed positive integer and for $1 \leqslant n \leqslant m$ let s_n be the number of surjections from $\{1, 2, \ldots, m\}$ to $\{1, 2, \ldots, n\}$. By considering the total number of functions from $\{1, 2, \ldots, m\}$ to $\{1, 2, \ldots, n\}$ show that for each n with

$1 \leqslant n \leqslant m$

$$\binom{n}{1}s_1 + \binom{n}{2}s_2 + \cdots + \binom{n}{n}s_n = n^m. \qquad [\text{H}]$$

Then use the method of inversion, from the previous exercise, to find a formula for s_n, thus confirming the result from exercise 15 on page 49. [A]

9. Let p and d be the generating functions associated with the partitions and distinct partitions, respectively, as on page 120 at the end of this chapter. Show that $p(x) = d(x)p(x^2)$ and deduce that

$$p_n = d_n + d_{n-2}p_1 + d_{n-4}p_2 + d_{n-6}p_3 + \cdots.$$

In addition give a direct argument to justify this last result.

10. In chapter 1 we observed that the number of shortest routes from A to B in the $n \times n$ grid shown is $\binom{2n}{n}$. Now let u_n be the number of shortest 'upper' routes from A to B; i.e. those which never go below the diagonal AB. In addition let v_n be the number of those upper routes which touch the diagonal AB only at A and B.

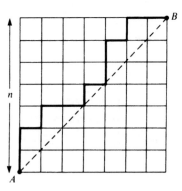

Show that $v_n = u_{n-1}$ for $n > 1$. Then, by considering the *last* point at which the route touches the diagonal, show that the u_ns satisfy the recurrence relation

$$u_0 = 1 \quad \text{and} \quad u_n = u_0 u_{n-1} + u_1 u_{n-2} + u_2 u_{n-3} + \cdots + u_{n-1} u_0 \quad (n > 0). \quad [\text{H}]$$

Let u be the generating function of the u_ns. Deduce that

$$x(u(x))^2 - u(x) + 1 = 0$$

and hence show that $u_n = \dfrac{1}{n+1}\binom{2n}{n}$ (these are the *Catalan numbers* already introduced briefly in the solution of exercise 13 of chapter 1). [H]

11. (i) For non-negative integers m and n let $s(m, n)$ be the coefficient of x^m in $x(x-1)(x-2)\cdots(x-n+1)$ (with the convention that $s(0, 0) = 1$). These are known as the *Stirling numbers of the first kind*: show that they are defined by the recurrence relation

$$s(0, 0) = 1, \qquad s(m, 0) = 0 \ (m > 0), \qquad s(0, n) = 0 \ (n > 0)$$

and

$$s(m, n) = s(m-1, n-1) - (m-1)s(m-1, n) \qquad (m, n > 0).$$

(ii) For non-negative integers m and n let $S(m, n)$ be the number of different partitions of the set $\{1, 2, \ldots, m\}$ into n non-empty sets (with the convention that

$S(0, 0) = 1$). These are known as the *Stirling numbers of the second kind*: show that they are determined by the recurrence relation

$$S(0, 0) = 1, \quad S(m, 0) = 0 \quad (m > 0), \quad S(0, n) = 0 \quad (n > 0)$$

and

$$S(m, n) = S(m - 1, n - 1) + nS(m - 1, n) \quad (m, n > 0).$$

(iii) Use the formula for the number of surjections from $\{1, 2, \ldots, m\}$ to $\{1, 2, \ldots, n\}$ (from exercise 8) to write down a specific formula for $S(m, n)$. [A]
(iv) Let M and N be the 4×4 matrices with (m, n)th entries $s(m, n)$ and $S(m, n)$ respectively ($1 \leq m, n \leq 4$). Calculate the entries of M and N and confirm that $M^{-1} = N$. [A]

(This works for any size matrices and is one of the key links between the two kinds of Stirling numbers. For this and further properties of these numbers – and for other enumeration problems in general – see an advanced text such as M. Aigner's *Combinatorial theory* or R.P. Stanley's *Enumerative combinatorics*.)

12. Use the method of generating functions to find two different six-sided dice with a positive integer on each face so that for any n the number of ways of obtaining a total of n, with the two dice is the same as with a traditional pair of dice. Which pair would you rather use for *Monopoly*? [H,A]

11

Vertex-colourings

Given a graph we now wish to colour its vertices so that no two vertices joined by an edge have the same colour (naturally enough called a *vertex-colouring*). For any particular graph what is the smallest number of colours needed?

Example

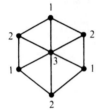

A vertex-colouring
in 3 colours, where
the 'colours' are 1, 2, 3

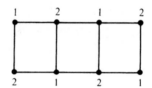

A vertex-colouring
in 2 colours, where
the 'colours' are 1 and 2   □

For a graph G the minimum number of colours needed in a vertex-colouring of it is called the *vertex-chromatic number* (or simply the *chromatic number*) of G and is denoted by $\chi(G)$: this is analogous to the edge-chromatic number $\chi_e(G)$ introduced in chapter 7.

Examples *It is clear that the complete graph K_n needs n colours in any vertex-colouring. Also it is easy to confirm that if a graph consists of a single cycle, then if the number of vertices is even two colours are sufficient to vertex-colour the graph, whereas if the number of vertices is odd three colours are needed.*

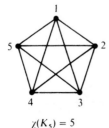
$\chi(K_5) = 5$

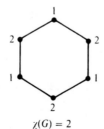
$\chi(G) = 2$

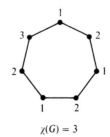
$\chi(G) = 3$ □

Let $G = (V, E)$ be a graph in which the highest degree of the vertices is d. As we saw in chapter 7, the edge-chromatic number satisfies the inequalities $d \leq \chi_e(G) \leq d + 1$. The first examples of this chapter show that the lower bound fails for the vertex-chromatic number $\chi(G)$ but, as we shall see in the exercises, it is very easy to show that $\chi(G)$ satisfies the inequality $\chi(G) \leq d + 1$. Indeed R.L. Brooks showed in 1941 precisely which connected graphs have $\chi(G) = d + 1$ and we shall state and prove his theorem after the next example (we restrict attention to connected graphs, for otherwise we could simply colour each component independently).

In our work on edge-colourings we used the technique of reversing two colours in various subgraphs. Although there are similar proof techniques for vertex-colouring results, for variety we use instead an algorithmic approach: the following example illustrates the technique.

Example *Let $G = (V, E)$ be as shown (with highest vertex degree $d = 4$):*

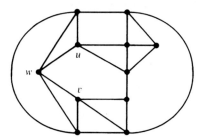

The following process shows how to vertex-colour G in four colours.

(i) *The vertices labelled u, v, w have $uw, vw \in E$, $uv \notin E$ and are such that the removal of u and v (and all edges ending there) will leave a connected graph G'. For each vertex x of G' find a 'shortest' path from w to x in G'; i.e. a path which uses as few edges as possible. The number of edges in this path is the 'distance' of x from w in G', and the distance of each vertex from w is noted below:*

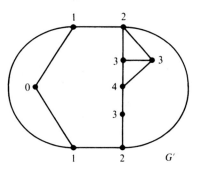

(*Note that each vertex of distance α (>0) from w is joined to a vertex of distance $\alpha - 1$ from w.*)

(ii) *Label the vertices of G in the following way. Let $v_1 = u$, $v_2 = v$, let v_3 be the vertex 'furthest away' from w in G', and label the remaining vertices as*

$v_4, \ldots, v_{10}, v_{11} = w$ so that they get progressively closer to w (i.e. so that the distance of v_5 from w in G' is less than or equal to the distance of v_4 from w, etc.):

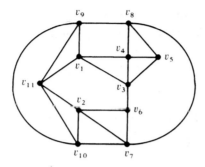

(*Note that this process ensures that each vertex v_i ($i < 11$) is joined to some higher-numbered vertex.*)

(iii) *Let the 'colours' be numbered 1, 2, 3, 4. Colour the vertices of G in the order $v_1, v_2, \ldots, v_{10}, v_{11}$ such that at each stage the colour used for vertex v_i is the lowest-numbered colour not yet used at a vertex joined to v_i:*

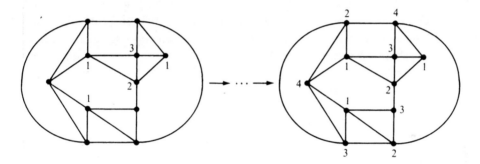

In this way we obtain the vertex-colouring of G in the four colours as shown. This process works because when colouring v_i ($1 \leq i < 11$) it is joined to a higher-numbered (and hence uncoloured) vertex: it is therefore joined to at most $d - 1$ ($= 3$) coloured vertices and one of the four colours will be available for it. And when colouring v_{11} it is joined to two vertices (v_1 and v_2) of colour 1 so there will again be a spare colour available for it. □

Theorem (Brooks) *Let $G = (V, E)$ be a connected graph with highest vertex degree d. Then if G is a complete graph, or if it consists of a single cycle with an odd number of edges, it follows that $\chi(G) = d + 1$. In all other cases $\chi(G) \leq d$.*

Proof The result is trivial for complete graphs or for cycles of any size. So we assume that G is neither complete nor a cycle and we show that it can be vertex-coloured in d colours. The proof is by induction on the number of vertices, the smallest proper case of $|V| = 3$ being trivial to check. So assume that G is as stated, that $|V| > 3$ and that the theorem is known for graphs of fewer vertices.

If G has a vertex w whose removal disconnects it, then the induction hypothesis can be applied to each of the graphs G_i (as illustrated below). It is soon clear that each G_i can be vertex-coloured in the same d colours and with the vertex w the same colour in each case. The various G_is can then be put together in the obvious way to give the required vertex-colouring of G in d colours.

So now assume in addition that G has no vertex whose removal will disconnect it. We saw in exercise 12 on page 80 that the complete graphs and cycles are characterised by the facts that no single vertex disconnects them but if u, v and w are vertices with $uw, vw \in E$ and $uv \notin E$ then the removal of u and v does disconnect them. (Details of the proof can be found in the solutions on page 232.) Since G is neither a cycle nor a complete graph, and since no single vertex disconnects it, in this case there must exist vertices u, v and w with $uw, vw \in E$, $uv \notin E$ and such that the removal of the vertices u and v gives a connected graph G'.

As in the above example label the vertices of G as $v_1 = u$, $v_2 = v$, and $v_3, \ldots, v_{n-1}, v_n = w$ in decreasing order of their distance from w in G'. This labelling then has the key property that each v_i ($1 \leqslant i < n$) is joined to some higher-numbered vertex. This is immediate for $i = 1$ or $i = 2$, and for $i > 2$ if you consider a shortest path from v_i to w in G' then it must be of the form

$$v_i, \quad v_j \quad , \ldots, w$$
$$\text{distance from } w: \quad \alpha \quad \alpha - 1 \quad \quad 0$$

which means that $v_i v_j \in E$ and that $j > i$.

So now let the 'colours' be numbered $1, 2, \ldots, d$ and colour the vertices of G in the order $v_1, v_2, \ldots, v_{n-1}, v_n$ such that at each stage the colour used for vertex v_i is the lowest-numbered colour not yet used at a vertex joined to v_i. Then, as in the example, v_1 and v_2 will both have colour 1. Also, when colouring v_i ($3 \leqslant i < n$) it will be joined to at least one higher-numbered (and hence uncoloured) vertex: as it has degree at most d it will therefore be joined to at most $d - 1$ coloured vertices. Hence there will always be one of the d colours available for $v_1, v_2, \ldots, v_{n-1}$. Finally, as v_n is joined to u and v, both of which have colour 1, there will also be a spare colour available for v_n. So in this way we will obtain a vertex-colouring of G in d colours as required, and the proof by induction is complete. □

Vertex-colourings

The next question we ask is not how many colours are needed for a vertex-colouring but, given some specific colours, in how many different ways the vertex-colouring can be done.

Example *Let G be the graph shown. Illustrate the number of different ways of vertex-colouring G with three colours 1, 2 and 3.*

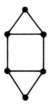

Find a formula for the number of ways of vertex-colouring G in k given colours.

Solution It is very straightforward to find the different vertex-colourings of G in the three given colours; there are 18 ways in all, as illustrated:

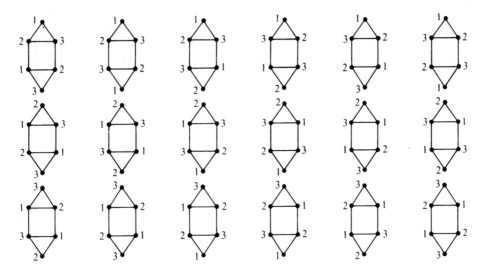

To find a formula for the number of ways of vertex-colouring G in k given colours it is convenient to distinguish two types of colourings, the first type in which the vertices u and v (as shown on the right) are of the same colour and the second type where u and v are of different colours. For this particular graph it is then easy to count the number of vertex-colourings of each type and to add the two totals together:

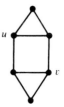

u and v of different colours

Let G_1 be the graph obtained from G by adding the edge uv. Then vertex-colouring G with u and v of different colours is clearly equivalent to vertex-colouring G_1. We now imagine vertex-colouring G_1 in any of k available colours, seeing how many choices we have at each stage. For this simple situation let us imagine colouring the vertices in order from left to right, noting the number of choices of colour available at each stage:

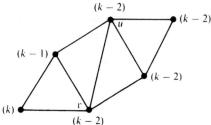

For the left-hand vertex there are k colours available, but then the next must be a different colour and so there are only $k - 1$ colours available for it. The next vertex (being joined to two vertices of different colours) has $k - 2$ colours available for it, and so on. Hence the total number of different ways of vertex-colouring G_1 (i.e. of vertex-colouring G with u and v of different colours) is

$$k(k-1)(k-2)^4.$$

u and v the same colour

Let G_2 be the graph obtained from G by 'contracting together' or 'merging' the two vertices u and v; i.e. they become a single vertex joined to any vertex which either one of them was joined to previously:

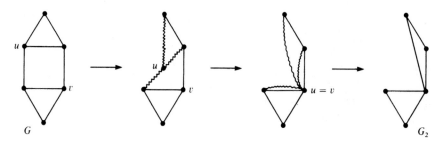

Then a little thought shows that vertex-colouring G with u and v the same colour is equivalent to vertex-colouring the very simple graph G_2. To see how many ways in which this can be done with k available colours, we again colour the vertices of G_2 from left to right noting the number of choices of colour available at each stage:

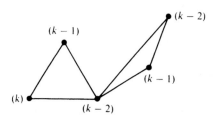

Vertex-colourings

Hence the total number of different ways of vertex-colouring G_2 (i.e. of vertex-colouring G with u and v the same colour) is

$$k(k-1)^2(k-2)^2.$$

So the total number of ways of vertex-colouring G is the sum of the number of ways in which u and v are different colours with the number of ways in which u and v are the same colour; i.e.

$$k(k-1)(k-2)^4 + k(k-1)^2(k-2)^2 = k(k-1)(k-2)^2((k-2)^2 + (k-1))$$
$$= k(k-1)(k-2)^2(k^2 - 3k + 3).$$

Note that when $k = 3$ this gives the answer 18, agreeing with the first part of the solution. □

In that example the number of ways of vertex-colouring G in k colours was a polynomial in k, and that will be true for any graph, as we now see.

Theorem *Given any graph G and any positive integer k, let $p_G(k)$ denote the number of different ways of vertex-colouring G using any of k available colours. Then p_G is a polynomial in k (known as the* chromatic polynomial *of G).*

Proof The proof is by induction on m, the number of 'missing edges' of G; i.e. on the number of additional edges needed to make $G = (V, E)$ into a complete graph on vertex-set V. In the case $m = 0$ the graph G must itself be a complete graph and, as each vertex must be of a different colour, it is clear that

$$p_G(k) = k(k-1)(k-2)\cdots(k-|V|+1),$$

which *is* a polynomial in k.

So now assume that $m > 0$ and that the result is known for graphs of fewer missing edges. Then G has at least one missing edge so let u and v be a pair of unjoined vertices of G. Let G_1 be the graph obtained from G by adding the edge uv and let G_2 be the graph obtained from G by 'contracting together' the vertices u and v; i.e. they become a single vertex joined to any vertex which was joined to either u or v in G. Then, as in the previous example, it is clear that

$p_G(k)$ = number of ways of vertex-colouring G in k colours

$$= \begin{pmatrix} \text{number of ways of vertex-} \\ \text{colouring } G \text{ in } k \text{ colours} \\ \text{with } u \text{ and } v \text{ different} \end{pmatrix} + \begin{pmatrix} \text{number of ways of vertex-} \\ \text{colouring } G \text{ in } k \text{ colours} \\ \text{with } u \text{ and } v \text{ the same} \end{pmatrix}$$

$$= \begin{pmatrix} \text{number of ways of vertex-} \\ \text{colouring } G_1 \text{ in } k \text{ colours} \end{pmatrix} + \begin{pmatrix} \text{number of ways of vertex-} \\ \text{colouring } G_2 \text{ in } k \text{ colours} \end{pmatrix}$$

$$= p_{G_1}(k) + p_{G_2}(k).$$

But each of G_1 and G_2 has fewer missing edges than G and so, by the induction hypothesis, both $p_{G_1}(k)$ and $p_{G_2}(k)$ are polynomials in k. Hence their sum, $p_G(k)$, is also a polynomial in k and the proof by induction is complete. □

In that proof we showed that the chromatic polynomial p_G satisfies the recurrence relation $p_G = p_{G_1} + p_{G_2}$, and that relation can be used as the basis of a neat algorithm to actually find p_G. Given a graph G we keep reapplying the recurrence relation until complete graphs are reached. Then the chromatic polynomial of G is simply the sum of the chromatic polynomials of all the complete graphs obtained.

Example *Find the chromatic polynomial of the graph G illustrated:*

Solution We start with G, choose an unjoined pair of vertices u and v, and then illustrate G_1 to the left below G and G_2 to the right below G. Then for each of the new graphs we choose a new pair of unjoined vertices and repeat the process: we continue until complete graphs are reached.

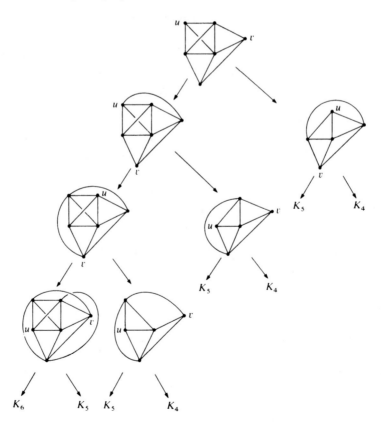

Hence

$$p_G(k) = p_{K_6}(k) + 4p_{K_5}(k) + 3p_{K_4}(k)$$
$$= k(k-1)(k-2)(k-3)(k-4)(k-5) + 4k(k-1)(k-2)(k-3)(k-4)$$
$$+ 3k(k-1)(k-2)(k-3)$$
$$= k(k-1)(k-2)(k-3)(k^2 - 5k + 7). \qquad \square$$

Exercises

1. Show that a graph $G = (V, E)$ is bipartite if and only if $\chi(G) \leq 2$.

2. Given an integer $d > 1$ illustrate a graph G with highest vertex degree d and with $\chi(G) = 2$.

3. Let G be a graph. Show that it has at least $\chi(G)$ vertices of degree $\chi(G) - 1$ or more. [H]

4. Without using Brooks' theorem, show by induction on the number of vertices that a graph with highest vertex degree d can be vertex-coloured in $d + 1$ colours.

5. Let G be a connected graph with highest vertex degree d which has a vertex w of degree less than d. By labelling the vertices in decreasing order of their distance from w, describe a simple algorithm for vertex-colouring G in d colours. [A]

6. Let $G = (V, E)$ be a graph and let $\bar{G}$ be its complement (as defined in exercise 4 on page 79). Show that $\chi(G) \cdot \chi(\bar{G}) \geq |V|$ and that $\chi(G) + \chi(\bar{G}) \leq |V| + 1$. [H]

7. Let G be a graph and let n be a positive integer. Show that $(x - n)$ is a factor of the polynomial $p_G(x)$ if and only if $n < \chi(G)$.

8. Use the method given in the last example of the chapter to find the chromatic polynomial of the graph G illustrated:

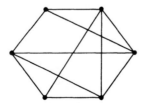

[A]

9. Find the chromatic polynomial of the graph illustrated:

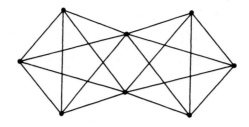

[H,A]

10. Let G_1 be the graph obtained from K_n by removing two edges with no endpoint in common, and let G_2 be the graph obtained from K_n by removing two edges with a common endpoint. Show that

$$p_{G_1}(k) - p_{G_2}(k) = k(k-1)(k-2)\cdots(k-n+3).$$ [H]

11. (i) Let G_1 and G_2 be two graphs with a single vertex in common and let the graph G be their 'union', as illustrated:

 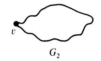 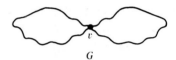

G_1 G_2 G

Show that $p_G(k) = \dfrac{1}{k} p_{G_1}(k) \cdot p_{G_2}(k)$.

(ii) In how many ways can the following graph be vertex-coloured in k colours? In how many of those colourings will $v_0, v_1, \ldots, v_n$ all be the same colour?

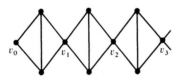 $\cdots$

[A]

12. Show by induction on n that the graph illustrated, with $2n$ vertices, has chromatic polynomial $k(k-1)(k^2 - 3k + 3)^{n-1}$:

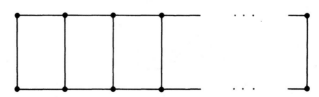

[H]

13. (i) Let G be a graph and let $e = uv$ be an edge of G. Let G_1 be the graph obtained from G by removing e, and let G_2 be the graph obtained from G by removing e and contracting together the vertices u and v. Show that

$$p_G(k) = p_{G_1}(k) - p_{G_2}(k).$$ [H]

(ii) Deduce that a tree on n vertices has chromatic polynomial $k(k-1)^{n-1}$. [H]

(iii) Find the smallest pair of non-isomorphic graphs with the same chromatic polynomial. [A]

14. Show that for a graph $G = (V, E)$ the chromatic polynomial is of the form

$$p_G(k) = k^{|V|} - |E|k^{|V|-1} + \text{lower powers of } k.$$ [H]

15. (*Stanley's theorem*, 1972.) An *acyclic orientation* of a graph $G = (V, E)$ is an ordering of each of its edges uv either as $u \to v$ or as $v \to u$ so that there is no cycle $v_1, v_2, \ldots, v_n, v_1$ with $v_1 \to v_2 \to \cdots \to v_n \to v_1$. (This is equivalent to putting an arrow along each edge whilst producing no directed cycle.)

Let $G = (V, E)$ have chromatic polynomial p_G. Show that G has exactly $(-1)^{|V|} p_G(-1)$ acyclic orientations. [H]

12

Rook polynomials

In this chapter we shall see how several different types of problems can be reduced to ones involving placing 'rooks' (or 'castles') on a chess-board.

Example *Four people A, B, C and D are to be given jobs a, b, c and d so that each of them gets one job. Person A will not do jobs b or c, person B will not do job a, person C will not do jobs a, b or d, and person D will not do jobs c or d. In how many different ways can the four jobs be allocated?*

Solution For the moment we will not actually solve this problem but we will show how it can be expressed as a problem involving rooks on a board. Consider the chart of people/jobs illustrated:

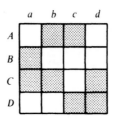

If we allocate job c to person B, for example, then we shall put a tick in the square corresponding to B/c. We have therefore blacked out those squares which correspond to an allocation of a job to an inappropriate person. To allocate all four jobs is now equivalent to placing four ticks in the unshaded part of the chart with no two ticks in the same row or column.

We are now ready to rephrase the problem in terms of rooks on a chess-board. First of all think of the above picture as a small chess-board. In chess a rook can move along its own row or column (i.e. its own rank or file), so a collection of rooks on a chess-board will be 'non-challenging' if no two of them are in the same row or column. Therefore the above problem of job allocation is equivalent to placing four non-challenging rooks on the unshaded part of the board illustrated. One such way

Rook polynomials

is shown below: it is equivalent to the allocation of jobs *A/d*, *B/b*, *C/c* and *D/a*.

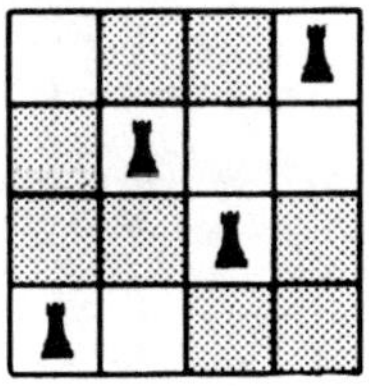

□

Example *How many permutations are there of* $\{1, 2, 3, 4, 5\}$ *in which no i is permed to i or i+1?*

Solution Again we shall not actually solve this yet but we will show how to transform the problem into one involving rooks. Consider the unshaded part of a 5×5 board shown below on which we wish to place five non-challenging rooks (i.e. with no two in the same row or column): one such arrangement is given.

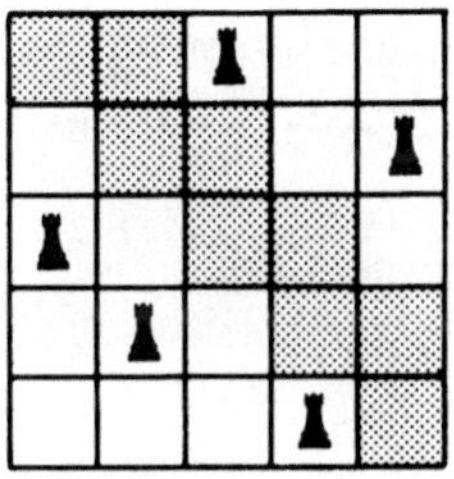

How is this related to permutations? If you think of the rook in row *i* and column *j* as meaning that *i* is mapped to *j* then the above arrangement corresponds to the mapping $1 \to 3$, $2 \to 5$, $3 \to 1$, $4 \to 2$ and $5 \to 4$. The fact that no two rooks are in the same row or column ensures that this mapping will be a permutation of $\{1, 2, 3, 4, 5\}$ and by deleting those particular squares we have made sure that no *i* gets permed to *i* or $i + 1$. So the required number of permutations is precisely the number of ways of placing five non-challenging rooks on the unshaded part of that 5×5 board.

□

Example *In how many ways can a fourth row be added to this* 3×5 *Latin rectangle to give a* 4×5 *Latin rectangle (with entries in* $\{1, 2, 3, 4, 5\}$*)?*

$$\begin{pmatrix} 4 & 1 & 5 & 3 & 2 \\ 1 & 2 & 4 & 5 & 3 \\ 2 & 5 & 3 & 4 & 1 \end{pmatrix}$$

Solution It is easy to see that adding a fourth row is equivalent to placing five non-challenging rooks on the unshaded board shown below; for example the arrangement illustrated corresponds to a new row of (3, 4, 2, 1, 5).

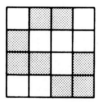

Those examples show how problems can be expressed in terms of rooks on a board. Imagine starting with a giant chess-board (of any finite size, $n \times n$ say) and then choosing a subset of its squares which we are allowed to use: these squares will form our 'board'. Given such a board, B, the *rook polynomial* of B is

$$r_B(x) = 1 + r_1 x + r_2 x^2 + r_3 x^3 + \cdots + r_k x^k + \cdots + r_n x^n$$

where r_k is the number of ways of placing k non-challenging rooks on B. (In fact the rook polynomial is another example of a generating function and, as before, we shall never be concerned with any particular numerical value of x.)

Examples *Find the rook polynomial of this 4×4 board from the job allocation problem which started the chapter:*

Solution We shall soon develop techniques for calculating rook polynomials but for the moment note that one rook can be placed on the board in eight different ways (i.e. there are eight available squares on the board). Then a laborious and careful count yields that two rooks (non-challenging, as always) can be placed on the board in 19 different ways, that three rooks can be placed on the board in 14 different ways, and four rooks in just two ways. So the board has rook polynomial

$$1 + 8x + 19x^2 + 14x^3 + 2x^4.$$

Changing one problem into another equivalent one does not necessarily make its solution any easier. But in the case of rook polynomials we shall use some of our earlier ideas (including recurrence relations and the inclusion/exclusion principle) to help us to calculate them easily. That will then enable us to solve the above exercises and a whole range of other combinatorial problems.

Rook polynomials

Our first labour-saving device concerns boards which can be split into two parts, neither of which has any effect upon the other.

Example *In the Latin rectangle example on the previous page we constructed the following board:*

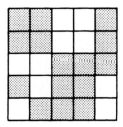

We leave it as a straightforward but laborious exercise to show that this board B has rook polynomial

$$r_B(x) = 1 + 10x + 35x^2 + 50x^3 + 26x^4 + 4x^5.$$

Now this particular board has the property that it can be cut into two boards C and D which have no row or column in common:

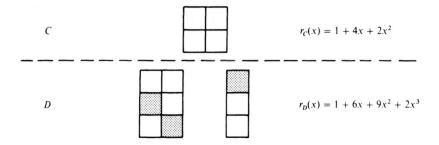

Note that

$$1 + 10x + 35x^2 + 50x^3 + 26x^4 + 4x^5 = (1 + 4x + 2x^2)(1 + 6x + 9x^2 + 2x^3),$$

i.e.

$$r_B(x) = r_C(x) \cdot r_D(x),$$

and this relationship will be true for any pair of non-interacting boards. □

Theorem *Let B be a board which can be partitioned into two parts C and D which have no row or column in common. Then*

$$r_B(x) = r_C(x) \cdot r_D(x).$$

Proof To prove the given identity we shall show that the coefficient of any x^k is the same on each side of the identity. Since the placing of rooks on C does not restrict the ways of placing rooks on D (and vice versa) it is easy to see that

coefficient of x^k in $r_B(x)$ = the number of ways of placing k rooks on B
= (the no. of ways of placing 0 rooks on C and k on D)
+ (the no. of ways of placing 1 rook on C and $k-1$ on D)
+ (the no. of ways of placing 2 rooks on C and $k-2$ on D)
$\vdots$
+ (the no. of ways of placing k rooks on C and 0 on D)
= (coeff. of x^0 in $r_C(x)$)·(coeff. of x^k in $r_D(x)$)
+ (coeff. of x^1 in $r_C(x)$)·(coeff. of x^{k-1} in $r_D(x)$)
+ (coeff. of x^2 in $r_C(x)$)·(coeff. of x^{k-2} in $r_D(x)$)
$\vdots$
+ (coeff. of x^k in $r_C(x)$)·(coeff. of x^0 in $r_D(x)$)
= the coefficient of x^k in $r_C(x) \cdot r_D(x)$,

and the theorem follows. $\square$

Example *The board of the type shown, but consisting of n unshaded squares, has rook polynomial $(1 + x)^n$.*

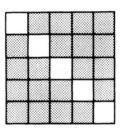

This can easily be checked from basics or proved by induction or, less formally, by noting that the board consists of n tiny 1×1 boards none of which has any row or column in common with any other. Each of these little boards has rook polynomial $1 + x$ and so by a natural extension of the previous theorem the given board's rook polynomial is the product of the n separate polynomials. $\square$

When finding the chromatic polynomial of a graph in the previous chapter we constructed two graphs from it and tried instead to find their chromatic polynomials (if this was still too hard we repeated the process). In this way we were using an algorithmic approach based on a recurrence relation. We now develop a not dissimilar process for finding the rook polynomial of a board.

Theorem *Let B be a board and let s be one particular square of that board. Then let B_1 be the board obtained from B by deleting the square s and let B_2 be the board obtained from B by deleting the whole row and column containing s. Then*

$$r_B(x) = r_{B_1}(x) + xr_{B_2}(x).$$

Rook polynomials

Proof As before, to establish the given identity we shall show that the coefficient of any x^k is the same on both sides of the identity: each of the following steps can easily be justified by the reader.

The coefficient of x^k in $r_B(x)$ = the number of ways of placing k rooks on B
$\qquad$ = (no. of ways of placing k rooks on B with s not used)
$\qquad$ + (no. of ways of placing k rooks on B with s used)
$\qquad$ = (no. of ways of placing k rooks on B_1)
$\qquad$ + (no. of ways of placing $k-1$ rooks on B_2)
$\qquad$ = (the coefficient of x^k in $r_{B_1}(x)$)
$\qquad$ + (the coefficient of x^{k-1} in $r_{B_2}(x)$)
$\qquad$ = (the coefficient of x^k in $r_{B_1}(x)$)
$\qquad$ + (the coefficient of x^k in $xr_{B_2}(x)$)
$\qquad$ = the coefficient of x^k in $r_{B_1}(x) + xr_{B_2}(x)$.

Hence $r_B(x) = r_{B_1}(x) + xr_{B_2}(x)$ and the theorem is proved. $\square$

Example *Before developing one further technique let us use the two previous theorems to calculate once again the rook polynomial of the board B of the job allocation problem which started with this chapter. At each stage choose the square s and draw B_1 on the left below and B_2 (with an extra factor x) on the right below.*

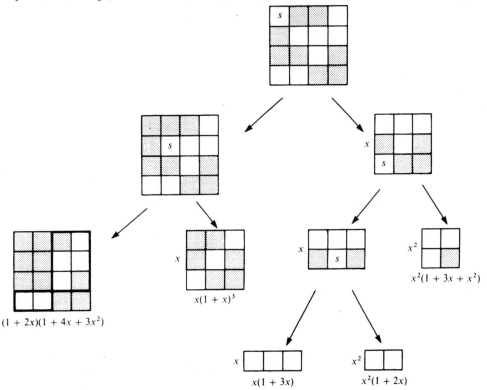

Hence

$$r_B(x) = (1 + 2x)(1 + 4x + 3x^2) + x(1 + x)^3 + x(1 + 3x)$$
$$+ x^2(1 + 2x) + x^2(1 + 3x + x^2)$$
$$= 1 + 8x + 19x^2 + 14x^3 + 2x^4,$$

as before. □

As that example illustrates, the two previous theorems combine to give us a reasonably neat technique for calculating any rook polynomial. However, those results are rather mundane and by themselves they are not enough to enable us to solve very neatly any problems which we could not solve previously. We now introduce a third result concerning rook polynomials which is much more interesting and which will enable us to solve some quite difficult problems.

Example *The board B on the left below is one which we constructed for the Latin rectangle example on pages 139–40. The board $\bar{B}$ on the right below is the 'complement' of B in the 5×5 board; i.e. $\bar{B}$ consists of the shaded squares of B. The rook polynomials of B and $\bar{B}$ are as stated ($r_B(x)$ was calculated earlier; the very keen reader can check that $r_{\bar{B}}(x)$ is correct).*

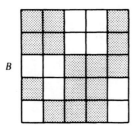

$r_B(x) = 1 + 10x + 35x^2 + 50x^3 + 26x^4 + 4x^5$

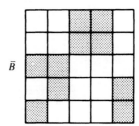

$r_{\bar{B}}(x) = 1 + 15x + 75x^2 + 145x^3 + 96x^4 + 12x^5$

It turns out that the coefficients of those two polynomials are interrelated in a far-from-obvious way:

coefficients in $r_B(x)$:   1,   10,   35,   50,   26,   4.

Note that $5! \times 1 - 4! \times 10 + 3! \times 35 - 2! \times 50 + 1! \times 26 - 0! \times 4$

$= 120 - 240 + 210 - 100 + 26 - 4$

$= 12$

which is the coefficient of x^5 in $r_{\bar{B}}(x)$!

Similarly,

coefficients in $r_{\bar{B}}(x)$: 1, 15, 75, 145, 96, 12.

and $5! \times 1 - 4! \times 15 + 3! \times 75 - 2! \times 145 + 1! \times 96 - 0! \times 12$

$= 120 - 360 + 450 - 290 + 96 - 12$

$= 4$

which is the coefficient of x^5 in $r_B(x)$! □

Theorem *Let B be part of an $n \times n$ board with rook polynomial*

$$r_B(x) = 1 + r_1 x + r_2 x^2 + \cdots + r_n x^n$$

and let $\bar{B}$ be the complement of B in the $n \times n$ board. Then the number of ways of placing n non-challenging rooks on $\bar{B}$ equals

$$n! - (n-1)! r_1 + (n-2)! r_2 - \cdots + (-1)^n 0! r_n.$$

Proof We use the inclusion/exclusion principle from page 44. To apply that principle we need a set of objects and some properties which those objects may or may not have. Let the set of objects be the layouts of n rooks (in non-challenging positions) on the full $n \times n$ board: there are $n!$ of these and in each there is precisely one rook in each row. Then let the possible properties $1, 2, \ldots, n$ of these layouts be as follows:

property 1 is that the rook in the first row is in B;

property 2 is that the rook in the second row is in B;

⋮

property n is that the rook in the nth row is in B.

Then the number of ways of placing n (non-challenging) rooks on $\bar{B}$ is precisely the number of the $n!$ layouts which have none of these n properties. By the inclusion/exclusion principle this number is

$$n! - N(1) - N(2) - \cdots - N(n)$$
$$+ N(1,2) + N(1,3) + \cdots + N(n-1, n)$$
$$- N(1,2,3) - N(1,2,4) - \cdots - N(n-2, n-1, n)$$
⋮
$$+ (-1)^n N(1, 2, \ldots, n),$$

where, in the standard notation from our work on the inclusion/exclusion principle, $N(i_1, i_2, \ldots, i_r)$ denotes the number of layouts which have at least the r different properties $i_1, i_2, \ldots, i_r$.

Now how can we calculate $N(i_1, i_2, \ldots, i_r)$? Let us consider, for example, $N(1, 2, 3)$.

This equals the number of ways of placing the rooks in rows 1, 2 and 3 in B and then the remaining $n - 3$ rooks can be freely placed in the remaining rows and columns in $(n - 3)!$ ways. Hence

$$N(1, 2, 3) + N(1, 2, 4) + \cdots + N(n - 2, n - 1, n)$$
$$= (n - 3)! \times \text{(number of ways of placing rooks in rows 1, 2 and 3 in } B)$$
$$+ (n - 3)! \times \text{(number of ways of placing rooks in rows 1, 2 and 4 in } B)$$
$$\vdots$$
$$+ (n - 3)! \times \text{(number of ways of placing rooks in rows } n - 2, n - 1 \text{ and } n \text{ in } B)$$
$$= (n - 3)! \times \text{(number of ways of placing three rooks anywhere in } B)$$
$$= (n - 3)! r_3.$$

In a similar way

$$N(1, 2, 3, 4) + N(1, 2, 3, 5) + \cdots + N(n - 3, n - 2, n - 1, n) = (n - 4)! r_4,$$

etc. Therefore the number of ways of placing n rooks on $\bar{B}$ is given by

$$n! - N(1) - N(2) - \cdots - N(n) + N(1, 2) + \cdots + (-1)^n N(1, 2, \ldots, n)$$
$$= n! - (n - 1)! r_1 + (n - 2)! r_2 - \cdots + (-1)^n 0! r_n,$$

as required. □

Example *In chapter 5 we counted the number of derangements of $\{1, 2, \ldots, n\}$ (i.e. the permutations in which $1 \not\mapsto 1, 2 \not\mapsto 2, \ldots, n \not\mapsto n$). But now we derive the result using rook polynomials.*

The number of derangements of $\{1, 2, \ldots, n\}$ is the number of ways of placing n rooks on the following $n \times n$ board:

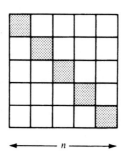

(Because, as before, if you think of a rook in the ith row and jth column as meaning that $i \to j$ then the excluded squares mean that there is a one-to-one correspondence between the layouts of n rooks on that board and the derangements of $\{1, 2, \ldots, n\}$.)

Rook polynomials

By the above theorem the number of ways of placing n rooks on the given board is

$$n! - (n-1)!r_1 + (n-2)!r_2 - \cdots + (-1)^n 0! r_n,$$

where

$$1 + r_1 x + r^2 x^2 + \cdots + r_n x^n$$

is the rook polynomial of the board

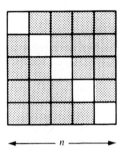

We saw earlier that the rook polynomial of this board is $(1 + x)^n$ and so $r_k = \binom{n}{k}$ and it follows that the number of derangements of $\{1, 2, \ldots, n\}$ is

$$n! - (n-1)!r_1 + (n-2)!r_2 - \cdots + (-1)^n 0! r_n$$

$$= n! - (n-1)!\binom{n}{1} + (n-2)!\binom{n}{2} - \cdots + (-1)^n 0!$$

$$= n! - \frac{n!}{1!} + \frac{n!}{2!} - \cdots + (-1)^n \frac{n!}{n!}$$

$$= n!\left(\frac{1}{2!} - \frac{1}{3!} + \cdots + (-1)^n \frac{1}{n!}\right),$$

as before. □

We are now able to use our results on rook polynomials to solve some interesting combinatorial problems. In order for the reader to derive the most benefit (and pleasure) from these solutions, the problems are left as exercises with plenty of guidance.

Exercises

1. Can a polynomial be both the rook polynomial of a board *and* the chromatic polynomial of a graph? [A]

2. (i) Find the rook polynomial of the following unshaded board:

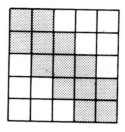
[A]

(ii) How many permutations are there of $\{1, 2, 3, 4, 5\}$ in which no i is permed to $i - 1$ or i or $i + 1$? How many are there in which each i is permed to $i - 1$ or i or $i + 1$? [A]

3. Use rook polynomials to find the number of orderings of 1, 2, 3, 4, 5 so that those numbers in that order are respectively members of the sets $\{2, 3, 4, 5\}$, $\{1, 2, 5\}$, $\{1, 4, 5\}$, $\{2, 3, 4\}$ and $\{1, 3, 4, 5\}$. [H, A]

4. Let

$$L = \begin{pmatrix} 1 & 2 & 3 & 4 & 5 \\ 4 & 3 & 5 & 1 & 2 \end{pmatrix}.$$

Use all three theorems from this chapter to calculate the number of ways of adding a third row to L to give a 3×5 Latin rectangle with entries in $\{1, 2, 3, 4, 5\}$. [H,A]

Confirm your answer by a direct count and deduce that L can be extended to a 5×5 Latin square in at least 24 different ways.

5. Consider an $n \times n$ board in which the squares are coloured black and white in the usual chequered fashion and with at least one white corner square.
(i) In how many ways can n non-challenging rooks be placed on the white squares? [A]
(ii) In how many ways can n non-challenging rooks be placed on the black squares? [A]

6. Let B be the board associated with the job allocation example which started this chapter. Verify that the complement of B in a 4×4 board has the same rook polynomial as B.

In general show that, if a board and its complement in an $n \times n$ board have the same rook polynomials, then n is even and the number of ways of placing 1 rook and $n - 1$ non-challenging rooks on the board are also even. [H]

7. Consider a board like the one illustrated but with n rows:

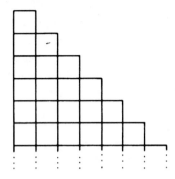

Show that its rook polynomial is

$$1 + S(n+1, n)x + S(n+1, n-1)x^2 + S(n+1, n-2)x^3 + \cdots + S(n+1, 1)x^n,$$

where the coefficients are Stirling numbers of the second kind, as defined in exercise 11 on pages 125–6. [H]

8. Let B be part of an $n \times n$ board with rook polynomial

$$r_B(x) = 1 + r_1 x + r_2 x^2 + \cdots + r_n x^n.$$

Also, for $0 \leq k \leq n$ let b_k denote the number of ways of placing n non-challenging rooks on the full $n \times n$ board with precisely k of them lying in B. Show that for $0 \leq k \leq n$

$$\binom{k}{k} b_k + \binom{k+1}{k} b_{k+1} + \binom{k+2}{k} b_{k+2} + \cdots + \binom{n}{k} b_n = r_k (n-k)! \qquad [H]$$

Deduce that

$$n! + r_1(n-1)! x + r_2(n-2)! x^2 + \cdots + r_n 0! x^n$$
$$= b_0 + b_1(1+x) + b_2(1+x)^2 + \cdots + b_n(1+x)^n.$$

Use this result to give an alternative proof of the theorem concerning complementary boards (and in fact inversion can be used to express b_k in terms of $r_k, r_{k+1}, \ldots, r_n$).

9. (i) Write down the rook polynomial of a full $n \times n$ board. [A]
   (ii) In how many ways can eight non-challenging rooks be placed on the following unshaded board?

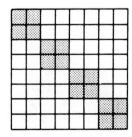

[A]

(iii) I have two packs of cards each consisting of two Jacks, two Queens, two Kings and two Aces. A game of 'snap' is as follows: each pack is shuffled and then the top card of each pack is compared, the second card of each is compared, and so on. A 'snap' is when two cards of the same denomination appear together. Find the probability that there is no snap in a game (i.e. find the proportion of all the possible orderings of the cards which result in no snaps: you could assume without any loss in generality that the order of the first pack is fixed).   [H,A]

(iv) What is the probability of no 'snap' when comparing two normal packs of 52 cards? (This is only for the keenest readers: the principles are the same as in (iii) but the final calculations need a computer – or a great deal of patience!)   [H,A]

10. (The hostess problem/*La problème des ménages*)

(i) Write down the rook polynomial of the board of the type shown but consisting of m squares:

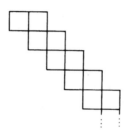

[H,A]

(ii) Find the rook polynomial of the *shaded* part of the following $n \times n$ board:

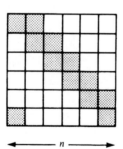

Hence find the number of ways of placing n non-challenging rooks on the unshaded board.   [H,A]

(iii) At a dinner-party n married couples are to be seated in the $2n$ seats around a circular table so that men and women alternate around the table and so that no man sits next to his own wife. In how many ways can this be done?   [H,A]

13
Planar graphs

We are familiar with the idea of illustrating a graph using some points in the plane to represent the vertices, and joining two of those 'vertices' by a line (or 'edge') whenever they are joined by an edge in the graph. Sometimes this can be done so that no two drawn edges meet except at a vertex which is an endpoint of them both.

Examples *The complete graph K_4 and the complete bipartite graph $K_{2,3}$ can both be drawn in that way:*

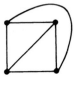

K_4

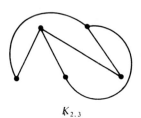

$K_{2,3}$

Attempting to draw K_5 and $K_{3,3}$ in a similar way seems to lead to problems:

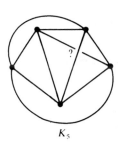

K_5

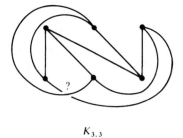

$K_{3,3}$

In fact we shall soon prove that it is impossible to draw these graphs in the plane without an extra intersection of edges. □

A graph $G = (V, E)$ is called *planar* if it can be drawn in the plane in the way just described; i.e. with no two edges meeting except at a common endpoint. Any such drawing of the graph is a *planar representation* of it. So the examples show that K_4 and $K_{2,3}$ are planar but we shall see later that K_5 and $K_{3,3}$ are not.

The idea of a planar graph, and other concepts which spring from that, bring together two areas of mathematics, graph theory (which is essentially the study of some relations) and the geometry of surfaces (which is part of topology). As often happens in mathematics, the marriage of two different disciplines results in some interesting (and sometimes rather difficult) offspring. In order to illustrate a wide range of interesting results in an elementary fashion some of the proofs in this chapter (and the next) will be rather sketchy or omitted altogether. In particular we shall deal with the geometrical/topological properties of the plane in a fairly casual and intuitive way.

We mention in passing that, although it is impossible to draw some graphs in the plane with no two edges meeting except at a common endpoint, it is always possible to illustrate any graph in three-dimensional space with this non-intersection property. Indeed, without going into full details, it is reasonably clear that any finite number of points can be chosen on the surface of a sphere so that no four of the points are coplanar. Using these points as the vertices of the graph and representing edges by straight line segments between appropriate pairs of vertices clearly gives a 'non-intersecting' straight line representation of the graph; i.e. with no two edges meeting except at a common endpoint *and* with each edge represented by a straight line.

Example *A 'non-intersecting' straight line representation of $K_{3,3}$ in three-dimensional space:*

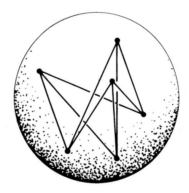

□

Given a planar representation of a graph we can consider the set of points in the plane not used by that representation: it will fall naturally into 'pieces' (formally they are the maximal topologically connected subsets) and they are called the *faces* of the representation.

Planar graphs

Example *The following figures are all planar representations of the same graph: in each case the faces have been numbered and face 1 is 'infinite', the rest are 'finite'.*

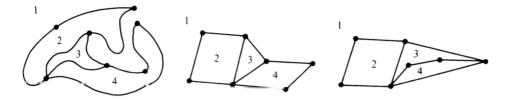

The first two representations are essentially the same (or 'equivalent') but the third is not because, for example, it has a face with a boundary consisting of five edges whereas the other representations do not. □

It is worth noting that, given a planar representation of a graph, it can be distorted to leave 'essentially the same' arrangement of vertices, edges and faces but with each edge now represented by a straight line. This amazing result was proved independently by K. Wagner in 1936 and I. Fáry in 1948. We state a slight extension of the result below and give the sketchiest of proofs. For the result we need the idea of a 'star-shaped' set: for the purposes of this result alone a set S in the plane is called *star-shaped* if there exists a circle C (of positive radius) within S such that for each point $x \in S$ and $y \in C$ the straight line segment from x to y lies entirely in S.

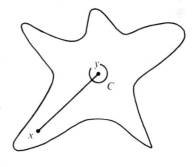

A star-shaped set S

Theorem (Fáry/Wagner) *Given any planar representation of a graph G and any particular finite face F whose boundary consists of a cycle, there exists an equivalent representation of G with each edge represented by a straight line segment and with the face F star-shaped.*

Sketch proof The proof is by induction on the number of vertices, the initial cases being trivial. So assume that we are given a planar representation of a graph G and a particular finite face F bounded by a cycle, and assume that the result holds for smaller graphs. We outline the deduction of the result for G in the case when there exists a vertex v on the boundary of F such that the removal of v and its edges from the planar representation leaves F as part of a larger finite face F' bounded by a cycle (as illustrated on the next page). By the induction hypothesis the reduced representation can be rearranged with each edge straight and with F' star-shaped. Since F' is star-shaped it is now possible to place v in F' so that v's edges can be added as non-intersecting straight lines, leaving F star-shaped, and giving the required representation of G:

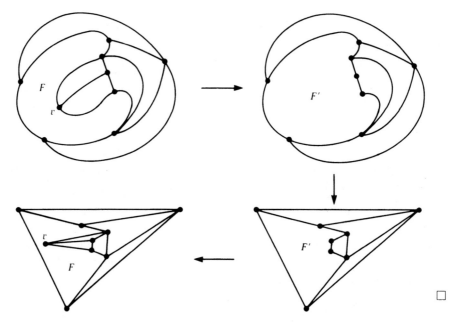

As we saw in the earlier example, there are many different planar representations of a particular planar graph. However, as happened in that example, they will all have the same number of faces, and we shall prove this in a moment. The result giving the number of faces in terms of the number of vertices and edges is known as *Euler's formula* because it is closely related to his result for the number of faces of a convex polyhedron. It was actually Cauchy in 1813 who essentially generalised Euler's result into one which incorporated planar graphs. (Fuller details of the historical development of the subject can be found in the excellent *Graph theory 1736–1936* mentioned several times before and listed in the bibliography.)

Example *The left-hand diagram shows a transparent solid cube. With the usual terminology for such solids, the cube has 8 vertices, 12 edges and 6 faces. The dodecahedron, shown on the right, has 20 vertices, 30 edges and 12 faces.*

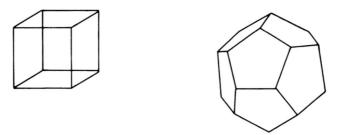

In each case

number of faces = number of edges − number of vertices + 2 and this was the general result for all such solids (*convex polyhedra*) given by Euler in 1752. Each of these solids can be opened up into a sort of 'map' of them in the plane, as we illustrate below. In each case we get a planar representation of a graph and the vertices/edges/faces of the solid become the vertices/edges/faces of the planar representation.

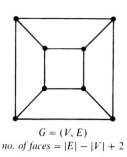

$G = (V, E)$
no. of faces $= |E| - |V| + 2$

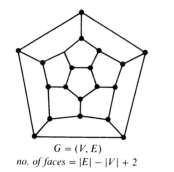

$G = (V, E)$
no. of faces $= |E| - |V| + 2$

Theorem (Euler's formula) Let $G = (V, E)$ be a connected planar graph. Then in any planar representation of G the number of faces will be $|E| - |V| + 2$.

Proof The proof is by induction on the number of cycles which G has. If it has no cycles then it is a tree and it is clear that any planar representation of G will have just one face. We know that for a tree $|E| = |V| - 1$ and so in this case

$$|E| - |V| + 2 = 1 = \text{the number of faces,}$$

as required. So assume that the planar graph G has at least one cycle and that the result is known for graphs of fewer cycles. Consider a planar representation of G. Let e be an edge of G which is part of one of its cycles and let $G' = (V, E\setminus\{e\})$ be the graph obtained from G by removing the edge e. Then deleting e from the given planar representation of G gives a planar representation of G'. Also G' is certainly connected and it has fewer cycles than G. Therefore we can apply the induction hypothesis to the representation of $G' = (V, E\setminus\{e\})$ to give that

number of faces of G' in this representation $= |E\setminus\{e\}| - |V| + 2 = |E| - |V| + 1$.

If we now replace the missing edge e in the representation how will this affect the number of faces? Since e was part of a cycle, replacing it will divide some face in the representation of G' into two faces, as illustrated below.

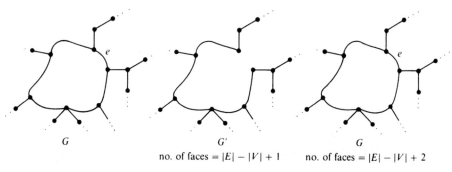

G

G'
no. of faces $= |E| - |V| + 1$

G
no. of faces $= |E| - |V| + 2$

Hence the given planar representation of G does have $|E| - |V| + 2$ faces as required, and the proof by induction is complete. □

First corollary *Let $G = (V, E)$ be a planar graph with k components. Then in any planar representation of G the number of faces is $|E| - |V| + k + 1$.*

Proof Given any planar representation of G the addition of some $k - 1$ edges will join up the components and give a planar representation of a connected planar graph $G' = (V, E')$ with the same number of faces as before; e.g.

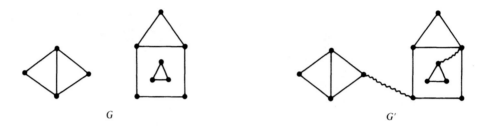

By Euler's formula the number of faces in the planar representation of G' (and therefore of G) is

$$|E'| - |V| + 2 = (|E| + k - 1) - |V| + 2 = |E| - |V| + k + 1,$$

as required. □

Note that no deduction from Euler's formula will enable us to show that any particular graph is planar because the theorem was of the form 'if G is planar **then** ...'. However it will enable us to deduce that certain graphs are *not* planar and we now have a string of further corollaries to the theorem for that purpose. In some of the proofs we want to count the edges used in the boundary of a face. There is no problem if the face is bounded by a cycle but in general if both sides of an edge are part of the boundary of the same face then that edge is counted twice in that face's boundary.

Example *In this example face 1 has a boundary of three edges, face 2 a boundary of six edges, and face 3 a boundary of eleven edges.*

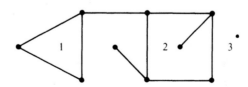

With that understanding the total of all the boundaries is 20 which is twice the number of edges. This will always be the case because each edge, having two 'sides', will be counted twice in total in all the boundaries. □

Planar graphs

Second corollary If $G = (V, E)$ is a planar graph with $|V| \geq 3$ then

$$|E| \leq 3|V| - 6.$$

Proof We can assume that G is connected (for otherwise we could add sufficient edges to make it connected and then deduce the inequality for the new graph and hence for G). Consider any planar representation of G and let f be the number of faces in that representation. The condition that $|V| \geq 3$ ensures that the boundary of each face must consist of at least three edges and so the total of all the boundary edges (counting an edge each time it occurs in a boundary) is at least $3f$. But, as remarked above, the total of all the boundary edges is precisely $2|E|$ since each edge will be counted twice in that grand total. Hence

$$2|E| \geq 3f = 3(|E| - |V| + 2)$$

and the required inequality follows. □

Example Show that the graph K_5 is not planar.

Solution If K_5 were planar, then by the previous corollary we would have

$$|E| \leq 3|V| - 6.$$

But for the complete graph on five vertices $|V| = 5$ and $|E| = 10$ and so $|E| \not\leq 3|V| - 6$. Hence K_5 is not planar. □

The next two corollaries virtually repeat the previous one but with added refinements.

Third corollary If $G = (V, E)$ is a planar bipartite graph with $|V| \geq 3$ then

$$|E| \leq 2|V| - 4.$$

Proof Recall first that a bipartite graph has no cycle consisting of an odd number of edges: it follows that the boundary of each face in the planar representation of a bipartite graph must consist of an even number of edges. Hence no boundary has just three edges and the boundary of each face consists of at least four edges. Therefore we can proceed exactly as in the proof of the second corollary but with this slight improvement:

$$\text{replace '3'} \qquad 2|E| \geq \underset{4}{\cancel{3}}(|E| - |V| + 2),$$
$$\text{by '4'}$$

from which the required inequality follows. □

Example *Show that the graph $K_{3,3}$ is not planar.*

Solution For this complete bipartite graph $|V| = 6$, $|E| = 9$ and so $|E| \not\leq 2|V| - 4$. Hence by the previous corollary $K_{3,3}$ is not planar. □

Fourth corollary *Let $G = (V, E)$ be a planar graph with a cycle and with r equal to the number of edges in its shortest cycle (this is called the* girth *of G). Then*

$$(r - 2)|E| \leq r(|V| - 2).$$

Proof In any planar representation of G each face has a boundary of at least r edges. Hence, by a further refinement of the second corollary,

$$2|E| \geq 3(|E| - |V| + 2),$$
$$r$$

from which the required inequality follows. □

Example *The* Petersen graph *(named after the nineteenth century Danish mathematician Julius Petersen) is as illustrated below. Show that it is not planar.*

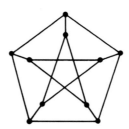

Solution For the Petersen graph the girth, r, is 5 (i.e. the shortest cycle has five edges). Also $|E| = 15$ and $|V| = 10$. Therefore

$$45 = (r - 2)|E| \not\leq r(|V| - 2) = 40$$

and it follows from the previous corollary that the Petersen graph is not planar. □

We have seen that K_5 and $K_{3,3}$ are not planar. In a remarkable theorem K. Kuratowski showed in 1930 that essentially these are the only two non-planar graphs, by proving that any non-planar graph must in some sense 'contain' K_5 or $K_{3,3}$.

Planar graphs

Example *We saw in the previous example that the Petersen graph is not planar: where within it can K_5 or $K_{3,3}$ be seen?*

The Petersen graph

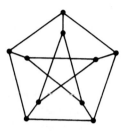

A subgraph of it

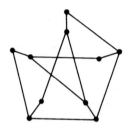

Vertices of degree 2 ignored

(i.e. regarded as stray vertices just resting on larger edges: they can just be blown away)

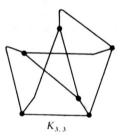

$K_{3,3}$

□

We must now make that idea of 'ignoring vertices of degree 2' more precise. Given an edge $e = uv$ of a graph, a *subdivision* of e is obtained by introducing a new vertex w and by replacing the edge uv with two edges uw and wv:

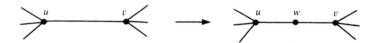

A *K-graph* is a graph which can be obtained from either K_5 or $K_{3,3}$ by a sequence of subdivisions. (Such graphs are also known as 'homeomorphic' to K_5 or $K_{3,3}$.)

Example *The following graphs are all K-graphs:*

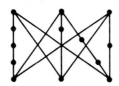  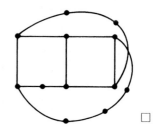

It is clear that a subdivision of an edge is just like adding a stray vertex onto it and that the planarity or otherwise of the graph will remain unaffected. Hence no K-graph will be planar. We now state the remarkable converse of this, namely that if a graph is not planar then it must contain a K-graph as a subgraph. The proof of this classic result can be found, for example, in J.A. Bondy and U.S.R. Murty's *Graph theory with applications*, but we merely illustrate one of the cases which arises in the proof.

Theorem (Kuratowski) *A graph is planar if and only if it contains no K-graph as a subgraph.*

Sketch proof By our comments above it only remains to show that any non-planar graph contains a K-graph. So the proof would consist of showing that if G is a minimal non-planar graph (in the sense that the removal of any edge will make it planar) and if G has no vertices of degree 2, then G is itself K_5 or $K_{3,3}$.

So let G be such a non-planar graph, let uv be an edge of G, and let G' be the graph obtained by deleting the edge uv from G. Then G' is planar and it can be shown that there is a cycle, C say, in G' which uses both u and v. In a planar representation of G' it is impossible to add the edge uv without crossing another edge, so in some way the representation of G' 'separates' u from v. The proofs of Kuratowski's theorem generally consider case by case the various ways in which this separation can be achieved.

The simplest case (which is the only one we shall consider) is when there is a path across the interior of C and a path across its exterior in the way illustrated:

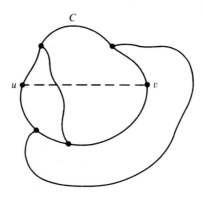

Planar graphs

But then the part of G illustrated is already a K-graph obtained by subdivisions from $K_{3,3}$ (in fact the minimality of G and the absence of vertices of degree 2 ensures that in this case the interior and exterior paths both consist of a single edge and that the above picture is the whole of G).

The various other ways in which u can be separated from v all eventually lead to a K_5 or $K_{3,3}$. □

We have seen that K_5 and $K_{3,3}$ are not planar, but we can adapt the figures from the first example of the chapter to draw each of these graphs with just one intersection of edges. So if we extend the surface of the plane by adding a little 'bridge' (topologists call them 'handles') then we can draw both these graphs without intersecting edges.

Example K_5 and $K_{3,3}$ *drawn without intersecting edges on a plane with one bridge/handle:*

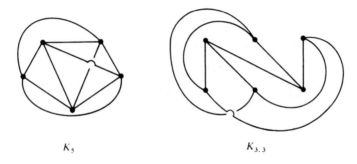

K_5 $\qquad$ $K_{3,3}$

Indeed that one extra bridge allows K_7 and $K_{4,4}$ to be drawn without intersecting edges: we illustrate K_6 and leave the others as exercises.

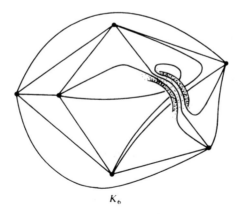

K_6 □

In general, if a minimum of g handles are needed to draw the graph G without intersecting edges, then G is said to be of *genus g*, written as $\gamma(G) = g$. Hence planar graphs have genus 0 and the above figure shows that the non-planar graph K_6 has genus 1.

It is possible to consider a representation of a graph on a surface consisting of the plane with additional handles/bridges and to obtain a version of Euler's formula for graphs of genus g. Some consequences of that formula (corresponding to our corollaries in the case of planar graphs) are now stated without proof.

Theorem *Let $G = (V, E)$ be a connected graph with $|V| \geqslant 3$. Then*

(i)
$$|E| - 6\gamma(G) \leqslant 3|V| - 6;$$

(ii) *if G is bipartite then*

$$|E| - 4\gamma(G) \leqslant 2|V| - 4;$$

(iii) *if G has girth r then*

$$(r - 2)|E| - 2r\gamma(G) \leqslant r(|V| - 2). \qquad \square$$

If we apply part (i) of that theorem to the complete graph K_n (with $|V| = n$ and $|E| = \frac{1}{2}n(n-1)$) then we deduce that

$$\gamma(K_n) \geqslant \tfrac{1}{12}n(n-1) - \tfrac{1}{2}n + 1 = \tfrac{1}{12}(n-3)(n-4).$$

Since the genus must be an integer it follows that

$$\gamma(K_n) \geqslant \{\tfrac{1}{12}(n-3)(n-4)\},$$

where $\{x\}$ denotes the smallest integer which is greater than or equal to x. So, for example, $\gamma(K_7) \geqslant \{1\} = 1$ and $\gamma(K_8) \geqslant \{\tfrac{5}{3}\} = 2$. In fact $\gamma(K_7) = 1$ and $\gamma(K_8) = 2$, and in general we have equality (as stated in the next theorem).

Similarly, part (ii) of the previous theorem can be applied to the complete bipartite graph $K_{m,n}$ to give

$$\gamma(K_{m,n}) \geqslant \{\tfrac{1}{4}(m-2)(n-2)\},$$

and once again equality holds.

Theorem (i) $\qquad\qquad \gamma(K_n) = \{\tfrac{1}{12}(n-3)(n-4)\},$

(ii) $\qquad\qquad \gamma(K_{m,n}) = \{\tfrac{1}{4}(m-2)(n-2)\},$

where $\{x\}$ denotes the smallest integer which is greater than or equal to x. $\qquad \square$

The result concerning the genus of the complete graphs was not proved until 1968 (by G. Ringel and J. Youngs): it is closely related to a problem in map-colouring, which is the topic of our next chapter.

Exercises

1. Let $G = (V, E)$ be a planar graph. Show that
   (i) if G has no vertices of degree 0 and G has a planar representation in which each face is bounded by three edges, then $|E| = 3|V| - 6$;
   (ii) If G has a planar representation in which no face is bounded by three or fewer edges, then $|E| \leq 2|V| - 4$.

2. Which of these graphs is planar? In the case of a planar graph draw a straight-line planar representation of it, and in the case of a non-planar graph find a K-graph which is a subgraph of it.

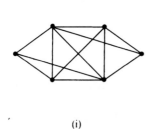
(i)

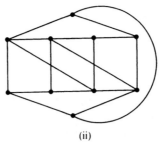
(ii)
[H,A]

3. Let G be a planar graph for which, in some planar representation, each face is bounded by a cycle containing an even number of edges. Show that G is bipartite. [H]

4. Illustrate each of K_7 and $K_{4,4}$ without intersecting edges on a plane with one handle (which is equivalent to drawing them on the surface of a 'torus', which is like a doughnut with a hole in the middle). [A]

5. The *thickness* t of the graph $G = (V, E)$ is the smallest number of planar graphs $(V, E_1), \ldots, (V, E_t)$ with $E_1 \cup \cdots \cup E_t = E$. Show that for $|V| \geq 3$ the thickness of G is at least $\{|E|/(3|V| - 6)\}$ and deduce that the thickness of the complete graph K_n is at least $[\frac{1}{6}(n + 7)]$.

6. Recall that the 'complement' of the graph $G = (V, E)$ is the graph $\bar{G} = (V, \bar{E})$ where

$$\bar{E} = \{vw \colon v, w \in V, v \neq w \text{ and } vw \notin E\}.$$

(i) Find a graph G with eight vertices so that G and $\bar{G}$ are both planar. (In fact it is possible to choose G so that, in addition, G and $\bar{G}$ are isomorphic.) [A]

(ii) Prove that if $|V| \geq 11$ then not both G and $\bar{G}$ can be planar. (In fact this result remains true for $|V| \geq 9$, but is very much harder to prove.) [H]

7. Consider an *n*-sided polygon with all of its chords added (as illustrated in the case $n = 6$) and assume that no more than two of these chords cross at any one intersection point. Make the figure into a planar graph by regarding each junction as a vertex (as shown). Then use Euler's formula to find the number of regions into which the polygon is divided. (This gives an alternative approach to exercise 12 of chapter 1.) [H,A]

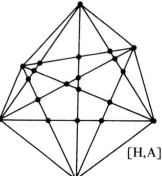

8. Let G be a planar graph in which each vertex has degree 3.
   (i) Show that, if G has a planar representation in which each face is bounded by precisely four or six edges, then there are exactly six faces bounded by four edges. [H]
   (ii) Show that, if G has a planar representation in which each face is bounded by precisely five or six edges, then there are exactly twelve faces bounded by five edges.
   In each case give two examples of such graphs with different numbers of vertices. [A]

9. Let G be a bipartite planar graph in which each vertex has degree d. Show that $d < 4$.
   Does there exist such a graph with $d = 3$? [A]

10. Let G be a planar graph with at least three vertices and for each non-negative integer n let v_n be the number of its vertices of degree n. Show that

$$\sum_{n=0}^{\infty} (6 - n)v_n \geq 12.$$ [H]

   Deduce that G has at least three vertices of degree 5 or less. [H]

11. Use exercise 10 to prove by induction that a planar graph can be vertex-coloured in six colours. [H]

12. Let $G = (V, E)$ be a connected planar graph in which each vertex has positive degree d and for which, in some planar representation, each face is bounded by c edges. Show that

$$\frac{1}{c} + \frac{1}{d} > \frac{1}{2}.$$ [H]

   List all pairs of integers c, d (≥ 3) which satisfy that inequality and in each case construct a graph G as described. [A]

13. *Sprouts* (or our slight variation of it) is a game for two players. Any number of vertices are marked in the plane. Then the two players take turns to add a new vertex and to join it by edges to two existing vertices. The only conditions are that the resulting picture must be a planar representation of a graph and that no vertex has degree more than 3. The loser is the first player who is unable to complete his/her turn. Show that, regardless of the number of initial vertices, the game must come to an end. [H]

14

Map-colourings

This chapter is concerned with colouring the regions of a map so that no two adjacent regions have the same colour.

Example

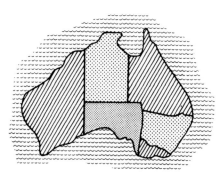

This map shows the mainland states of Australia in which each region (including the surrounding sea) has been 'coloured' with, of course, no two adjacent regions the same colour. We have used four 'colours' and, for this particular map, that is the lowest number of colours possible. □

How can map-colouring be treated mathematically as part of graph theory? There are two possible ways, as illustrated in the next example.

Example *The previous map of Australia can be regarded as a planar representation of a graph, as shown:*

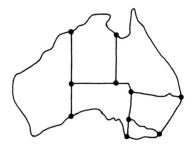

Map-colourings

A map-colouring is then merely a colouring of the faces of the planar representation so that no two faces which share a common boundary edge are the same colour.

Alternatively we can use a different graph to indicate which regions are adjacent. In the Australia example there will be seven vertices representing the seven regions and two vertices will be joined by an edge if their regions share one or more common boundaries. The resulting graph will in fact be planar because a planar representation of it can be superimposed on the original map, as shown:

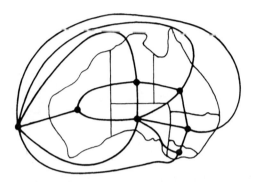

Colouring the original map is now equivalent to vertex-colouring this new graphical representation. □

That example shows that the colouring of a map can be regarded as colouring the faces of a planar representation of a graph or as vertex-colouring a new planar graph. For most of our results we will consider face-colourings, but we shall return to the vertex form later in the chapter.

When studying the face-colourings of planar representations of graph we shall restrict attention to graphs which can actually arise from sensible maps. Firstly we shall assume that the graphs are connected (for otherwise colouring the various components would be like colouring quite separate islands). Secondly we shall assume that the graph has no 'bridge'; i.e. no edge whose removal disconnects the graph. For in any planar representation of a graph with a bridge, that edge will have the same face on each side of it, and of course you never come across such boundaries in practice. So we shall now define a *map* as a planar representation of a connected planar graph with at least one edge but no bridges. Then each edge is part of the boundary of two different faces.

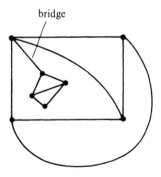

Now a *map-colouring* is obtained by assigning a colour to each face of the map so that no two faces with a common boundary edge have the same colour. The question which we address here is 'how many colours are needed to be able to colour every map?' The Australia example above needed four colours and, as you probably know, four colours will always be enough. Before discussing the extremely difficult 'four-colour theorem' we look at some lesser results concerning map-colouring. The first characterises those maps which can be coloured with just two colours.

Example *On the left below is a map with each vertex (of the graph associated with the map) of even degree. Hence the edges form a disjoint union of cycles, as illustrated in the middle figure. Using one colour for the faces inside an odd number of those cycles and a second colour for the faces inside an even number of those cycles gives the map-colouring shown on the right.*

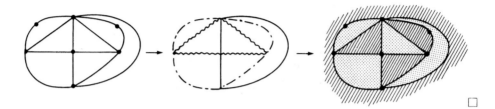

□

Theorem *A map can be coloured in two colours if and only if each vertex has even degree.*

Proof ($\Rightarrow$) Assume first that the map (with associated graph G) has been coloured in two colours, let v be any vertex of G, and let n be the degree of v. In the coloured map label the faces which meet at v as $f_1, f_2, \ldots, f_n$ clockwise around v, as illustrated. In fact these faces need not all be different; it is quite possible, for example, for faces f_1 and f_3 to be the same face. However, the absence of bridges means that f_1 is a different face from f_2, f_2 is different from $f_3, \ldots$, and f_n is different from f_1. Hence the two colours in the map alternate around v, as illustrated, and it follows that n is even. Therefore each vertex of G has even degree, as required.

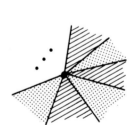

($\Leftarrow$) Conversely, assume that the graph G associated with the given map has each vertex of even degree. Then G is Eulerian and, by the theorem on page 72, G has some cycles $C_1, \ldots, C_r$ which between them use each edge of G exactly once. Now look at these cycles in the map: each has an 'inside', and for each face of the map we can count how many of these cycles it lies inside. If a face lies inside an odd number of the cycles $C_1, \ldots, C_r$ then use the first colour for that face, and if it lies inside an even number of the cycles then use the second colour for it (as in the previous example).

This certainly colours each face in one of two colours, but does it give a proper map-colouring, i.e. does each edge form part of the boundary of two faces of different colours? Let e be any edge of G: then e lies in just one of the cycles $C_1, \ldots, C_r$;

assume it is in C_i. Assume also that a face on one side of e has the first colour, say. Then that face lies inside an odd number of the cycles $C_1, \ldots, C_r$. Now crossing the edge e to get to the face on the other side means crossing the one cycle C_i. Hence it is easy to see that the face on the other side of e lies inside an even number of the cycles $C_1, \ldots, C_r$ and will be of the second colour. In this way we can see that we have coloured the map in two colours, as required. □

Have you ever seen a geographical map properly coloured just using two plain colours for the regions? It is highly unlikely because, as that theorem shows, each vertex would have even degree. Geographical maps do not give rise naturally to vertices of degree 2, and a vertex of 4 of more would mean at least four regions meeting at a point: do you know of a place where that happens? Certainly no four counties of Britain meet at a point (although Cambridgeshire, Leicestershire, Lincolnshire and Northamptonshire come very close to a point near Stamford). To mark the rare event of four regions meeting at a point, the junction of the four American states of Arizona, Utah, Colorado and New Mexico is aptly called Four Corners. In fact the vast majority of vertices in maps which arise from geography will have degree 3. For that reason we shall restrict attention to *cubic maps*, which are those in which each vertex has degree 3. There are also mathematical justifications for restricting attention to cubic maps.

Example *Suppose, for the moment, that Cambridgeshire, Leicestershire, Lincolnshire and Northamptonshire do in fact meet at a single point, v. Suppose also that you had a method for colouring all cubic maps in four colours but I then asked you to colour a map of the British counties in four colours. You could adapt the map as follows (it is a bit like re-inventing Rutland):*

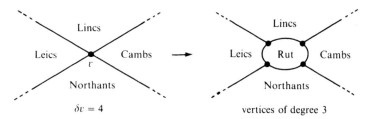

*Repeating that process at all vertices of degree greater than 3 would give a rather more complicated **cubic** map. You could then colour the new cubic map in four colours and then delete your newly invented counties to give a colouring of the original map in four colours:*

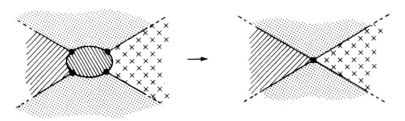

□

So, both geographically and mathematically, it is sensible to restrict attention to cubic maps. Before proceeding to theorems concerning the colouring of such maps we note, as a simple consequence of Euler's formula, that in any cubic map there will always be at least one face which is bounded by five or fewer edges. For otherwise each face would be bounded by six or more edges and we would have, by counting the boundary edges,

$$2|E| \leqslant 6(|E| - |V| + 2)$$

from which it would follow that $2|E| > 3|V|$. That would then contradict the fact that each vertex is of degree 3. Hence each cubic map does have a face bounded by five or fewer edges, and we can use that fact in the proofs of some of our colouring theorems. The next theorem answers the question of when three colours are sufficient for cubic maps.

Theorem *A cubic map can be coloured in three colours if and only if its graph is bipartite.*

Proof ($\Rightarrow$) Assume firstly that the cubic map (with associated graph G) has been coloured in three colours. We shall show that each cycle in G consists of an even number of edges which, by the theorem on page 42, will show that G is bipartite.

Before considering an arbitrary cycle of G let C_i be a cycle which forms the boundary of a face f_i of the map. Then, because each vertex has degree 3, the situation is as shown on the right: one of the colours will have been used for f_i and the other two colours will have to alternate around f_i. It follows that C_i has an even number of edges.

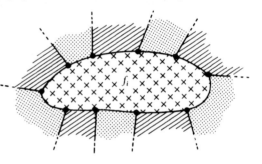

Now let C be any cycle of G and assume that in the map precisely the faces $f_1, \ldots, f_n$ lie inside C. Then (as in exercise 3 of the previous chapter) it follows that C itself contains an even number of edges. For if each f_i is bounded by the cycle C_i, as shown, and there are precisely m edges inside C then a simple inclusion/exclusion count gives

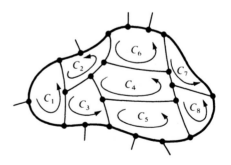

$$|C| = \underbrace{|C_1| + \cdots + |C_n|}_{\text{each is even}} - 2m = \text{even}.$$

Hence each cycle of G uses an even number of edges and G is bipartite, as required.

($\Leftarrow$) Conversely, assume that we are given a cubic map whose graph $G = (V, E)$ is bipartite. Then, since G is a bipartite planar graph, we have from the third corollary on page 157 that $|E| \leq 2|V| - 4$. Hence, using the fact that each vertex is of degree 3, we have

$$3|V| = 2|E| \leq 4|V| - 8$$

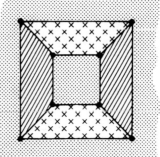

and it follows that $|V| \geq 8$ and that $|E| \geq 12$. In fact it is straightforward to confirm that the only cubic bipartite maps with $|E| = 12$ are all equivalent to the one shown on the right, and that can be coloured in three colours as shown.

The proof of the converse is now by induction on $|E|$, the lowest attainable case being $|E| = 12$, and we have just seen that the converse holds in that case. So assume that $|E| > 12$ and that the result holds for graphs of fewer edges. By the comments preceding this theorem the cubic map has a face bounded by three, four or five edges. But the bipartiteness of the graph means that no face is bounded by an odd number of edges and hence in this case the map has a face f bounded by precisely four edges. We illustrate one such face on the right and we have labelled the four adjacent faces f_1, f_2, f_3 and f_4 clockwise around f. It is quite possible that f_1 and f_3 are in fact part of the same face and it is also possible that f_2 and f_4 are part of the same face, but it is impossible for both those events to happen at the same time for that would involve the f_1/f_3 face crossing the f_2/f_4 face at some point. So let us assume without any loss in

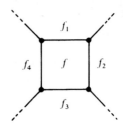

generality that f_1 and f_3 are different faces, as shown below. Then doctor f and its boundary, as illustrated, whilst leaving the rest of the map unchanged:

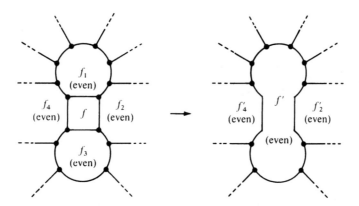

A few moments of thought will show you that the new figure is still a cubic map whose graph is bipartite: the fact that each vertex still has degree 3 is immediate and the fact that each face is still bounded by an even number of edges is easily checked. Furthermore the new map has fewer edges and so we can use the induction hypothesis to deduce that it can be coloured in three colours. Below we show one such colouring: notice that one colour is used for f' and the other two colours must alternate around f'. A simple parity check around the edges of f' shows that in any such colouring the faces f'_2 and f'_4 will have the same colour. It is then easy to reinstate the original map and to adapt the colouring for it without using any additional colours:

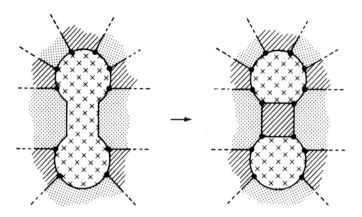

Hence we have deduced that the given map can be coloured in three colours, and the proof by induction is complete. □

Having considered two colours and three colours, we now move up to four. A little thought will show you that the previous theorem can be restated as saying that a cubic map can be coloured in three colours if and only if it can be *vertex*-coloured in two colours. The next result neatly ties in colouring a cubic map in four colours with *edge*-colouring it in three.

Example *Recall the earlier colouring of the map of Australia in four colours, and give the colours the names '00', '01', '10' and '11' as shown:*

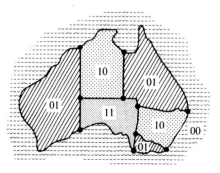

Map-colourings

Each edge borders two faces of different colours and we can use those colours (in a sort of binary additive way) to determine a colour for the edge:

e.g.   *10 on one side and 01 on the other* ⇒ *use 11 for the edge.*

Similarly,

$10/11 \Rightarrow 01$, $10/00 \Rightarrow 10$, $01/11 \Rightarrow 10$, $01/00 \Rightarrow 01$, $11/00 \Rightarrow 11$.

This gives the following edge-colouring of the graph in three colours:

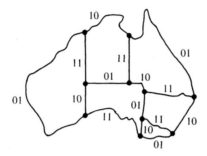

Suppose on the other hand that we had been given this edge-colouring: there is a neat way to construct from it a colouring of the map in four colours. In the left-hand figure below we have labelled with a '1' those faces which lie inside some cycle formed by edges of the two colours 10 and 11, the rest being labelled '0'; and on the right we have labelled with a '1' those faces which lie inside a cycle formed by edges of the two colours 01 and 11, the rest being labelled '0':

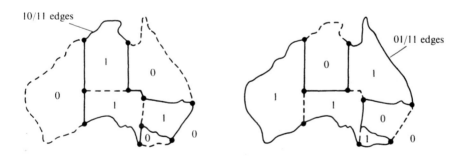

Those two pictures give each face two 'scores': for example Western Australia scored 0 on the left and 1 on the right and so we shall use the colour 01 for it. Overall not only does this give us a colouring of the map in the four colours 00, 01, 10 and 11, but it also in this case takes us back to the same colouring with which we started the example:

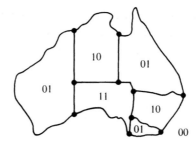

Theorem *A cubic map can be coloured in four colours if and only if its graph can be edge-coloured in three colours.*

Proof ($\Rightarrow$) Assume firstly that the cubic map (with associated graph G) has been coloured in four colours, and give the four colours the names '00', '01', '10' and '11'. As in the previous example we construct an edge-colouring of the graph in just three of these colours. To determine which colour to use for an edge, note the colours on each side of it and then use the following rules:

10 on one side and 01 on the other $\Rightarrow$ use 11 for the edge,

$10/11 \Rightarrow 01$, $\quad 10/00 \Rightarrow 10$, $\quad 01/11 \Rightarrow 10$, $\quad 01/00 \Rightarrow 01$, $\quad 11/00 \Rightarrow 11$.

This certainly colours each edge in one of the three colours 01, 10 and 11 but is it a proper edge-colouring; i.e. at each vertex do all the three edges meeting there have different colours? Suppose, for example, that faces of colours 00, 01 and 11 meet at a vertex: then the three edges meeting there will be coloured 01, 10 and 11 as shown:

The cases of the other combinations of three colours of faces meeting at a vertex can be checked similarly. Hence this construction does give an edge-colouring of the graph in three colours.

($\Leftarrow$) Conversely, assume that we are given an edge-colouring of the graph in the three colours 01, 10 and 11. Then each colour occurs at each vertex and so, in the subgraph formed by deleting the edges of colour 01, each vertex has degree 2. Hence each component of that subgraph consists simply of a cycle (and, although these are disjoint, it is quite possible, in the map, for one of these cycles to be inside another). Now in the map each face will be inside either an even number of these cycles (earning it a 'score' of 0) or an odd number (earning it a score of 1). Similarly, if instead we delete the 10 edges, then each face will be inside either an even number of the remaining cycles (earning it the second score of 0) or an odd number (earning it the second score of 1). Therefore this process gives each face one of the double scores 00, 01, 10 or 11 and, regarding these as colours, this does indeed give a coluring of the map in four colours. For it certainly colours each face and we must merely check that for any edge e the two faces on each side of e are of different colours.

Suppose, for example, that e is of colour 10 and that the face f on one side of it has colour 11. Then in the above process the face f has first score 1; i.e. it lies inside an odd number of cycles of edges of colours 10 and 11. But then crossing the edge e (of colour 10) will change that number of cycles by 1 and hence lead to a face f' in an even number of those cycles; i.e. having first score 0. The second score is unaffected in this case and so the colour of f' is 01 and the two faces on either side of e do indeed have different colours. In general two faces on either side of an edge of colour 10 have different first scores, two faces on either side of an edge of colour 01 have different second scores, and two faces on either side of an edge of colour 11 have different first scores and different second scores.

So in all cases an edge borders two faces of different colours and the four colours have indeed properly coloured the map: this completes the proof of the theorem.

□

Although that is a delightful result, it merely shows the equivalence of two very difficult properties. The first is that each cubic map (and hence each map) can be coloured in four colours: this is the famous 'four-colour theorem'. For the second property we recall our discussion of Vizing's theorem in chapter 7: if a graph has highest vertex-degree d then it can be edge-coloured in $d + 1$ colours. But for some graphs d colours will do, and it is in general a non-trivial problem to find simple categories of such graphs. The second property in the previous theorem says that all planar cubic graphs have this property.

The four-colour theorem has a long and interesting history, the early part of which is described in the much-referred-to *Graph theory 1736–1936*. The authors quote an 1852 letter from De Morgan as the earliest written reference to the problem and they discuss its development, including the famous fallacious 'proof' by A.B. Kempe in 1879 and P.J. Heawood's demonstration of the fallacy in 1890. This problem opened up many lines of study of graph theory, some of which we shall return to in a moment. But the four-colour theorem itself was not proved until 1976 when two American mathematicians, K. Appel and W. Haken, took up some of Kempe's ideas and produced a computer-aided proof: more details of their approach can be found in *Selected topics in graph theory* quoted in the bibliography. Appel estimated that their proof would take about 300 computer-hours and, although the new breed of machines will cut that time considerably, you will understand why there is no time for a proof in an introductory course! (We shall, however prove shortly that five colours are sufficient.)

Theorem (The four-colour theorem)   *Every map can be coloured in four colours.*   □

The many branches of the subject opened up by the study of map-colourings include edge-colouring results like the one proved earlier, 'maps' on other surfaces, and the study of 'duals' of maps where (as mentioned at the beginning of the chapter) the faces are replaced by vertices. In the case of 'maps' on other surfaces the 'Heawood map-colouring theorem' says that on a surface obtained from the plane by adding h handles any map can be coloured in $[\frac{1}{2}(7 + \sqrt{(1 + 48h)})]$ colours. For $h > 0$ the final piece of this proof was provided in 1968 when (as we discussed in the previous chapter) G. Ringel and J.W. Youngs finally proved that the complete graph K_n has genus $\{\frac{1}{12}(n - 3)(n - 4)\}$: for $h = 0$ the result is merely a restatement of the four-colour theorem.

In our proofs of map-colouring theorems we have taken the naive approach and considered colouring the faces of a map. In our final result in this chapter we are going to prove that all maps can be coloured in five colours, and we could prove this by similar techniques. We would use induction on the number of edges, choose a face bounded by three, four or five edges, and then in each of these cases reduce the map to a smaller one (as we did in one case in the proof on page 171). However the proof becomes a little less messy if we switch to a vertex-formulation of the problem: this will also enable us to illustrate an important general technique.

Given a map we can define a new planar graph with set of vertices equal to the set of faces of the map, and with two vertices joined by an edge if the two corresponding faces share one or more common boundaries. (If we wished to pursue this subject in more detail we would introduce graphs where repeated edges are allowed between a pair of vertices, we would let *each* edge of the map give rise to an edge of the new graph, and, in this way, lead to a study of 'dual' graphs.)

Example   *We recall the earlier Australia example and note that colouring the map of Australia is equivalent to vertex-colouring this new planar graph:*

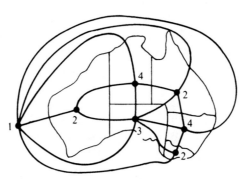

□

So, without going into too much detail, the fact that every map can be coloured in five colours is equivalent to the fact that any planar graph can be vertex-coloured in five colours, and we conclude this chapter with a proof of that result.

Theorem   *Every map can be coloured in five colours.*

Map-colourings

Proof As commented above, we shall prove the equivalent result that any planar graph $G = (V, E)$ can be vertex-coloured in five colours: we do this by induction on $|V|$, the cases $|V| \leq 5$ being trivial. So assume that $|V| > 5$, that we have a planar representation of G, and that planar graphs of fewer vertices can be vertex-coloured in five colours.

By the second corollary to Euler's formula (page 157) the planarity of G ensures that $|E| \leq 3|V| - 6$ and hence that

$$\sum_{v \in V} \delta v = 2|E| \leq 6|V| - 12 < 6|V|.$$

It follows that not each vertex of G can have degree 6 or more: so let v be a vertex of degree 5 or less. Then let G' be the graph obtained from G by the deletion of v (and the edges ending there). By the induction hypothesis we can vertex-colour G' in five colours: carry out one such colouring and then look at those colours (of all the vertices except v) in the planar representation of G.

If there is one of the five colours which is not used at a vertex joined to v, then clearly that colour can be used for v and it will extend the colouring to a vertex-colouring of G in five colours as required. So assume, on the other hand, that v is the endpoint of precisely the edges vv_1, vv_2, vv_3, vv_4 and vv_5 clockwise around v with the colour 1 used for v_1 and the colour 2 used for v_2 etc., as illustrated.

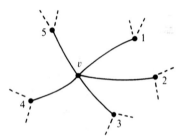

For each i and j with $1 \leq i < j \leq 5$, let G_{ij} be the subgraph of G' consisting of just those vertices of colour i and colour j and all the edges joining such vertices. If v_1 and v_3 are in different components of G_{13} then we can reverse the two colours in the component of G_{13} which contains v_1; i.e. in that component recolour in colour 3 all vertices which were originally colour 1, and vice versa. A little thought will show you that this still leaves a vertex-colouring of G' in the same five colours, but that now the colour 1 is not used at any vertex joined to v. Therefore we can colour the vertex v in that colour and obtain the required vertex-colouring of G.

A similar argument can be applied if v_2 and v_4 are in different components of G_{24}. So the only remaining case to consider is when v_1 and v_3 are in the same component of G_{13} and also v_2 and v_4 are in the same component of G_{24}. In this case there will be a path from v_1 to v_3 in which the vertices alternate in the colours 1/3 and there will also be a path from v_2 to v_4 in which the vertices alternate in the two colours 2/4. However, when we try to picture two such paths in the planar representation of G we see that such a situation is impossible. For the two paths would have to cross at some vertex; since that vertex is on both paths it would have to be an odd colour and an even colour at the same time:

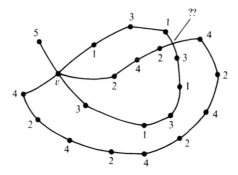

Hence we have already covered all possible cases and deduced that G can be vertex-coloured in five colours: this completes the proof by induction. □

Exercises

1. Consider a map with less than 12 faces and with the degree of each vertex at least 3. Show that there is a face bounded by four or fewer edges. [H]
   Without using the four-colour theorem, show that such a map can be coloured in four colours. [H]

2. Suppose that a collection of coins of the same size is spread out on a table so that no two coins overlap. We wish to colour the coins so that no two touching coins are the same colour. Without using the four-colour theorem, show that four colours are sufficient for any arrangement of coins, but that for some arrangements three colours are not enough. [H,A]

3. Show that a cubic map can be coloured in four colours if and only if it is possible to associate with each vertex v a number $p(v) = \pm 1$ so that for any cycle $v_1, v_2, \ldots, v_n, v_1$ whose edges form the boundary of a face we have
$$p(v_1) + \cdots + p(v_n) \equiv 0 \pmod{3}.$$
[H]

4. Given two graphs $G_1 = (V, E_1)$ and $G_2 = (V, E_2)$ let $G_1 \cup G_2$ be the graph $(V, E_1 \cup E_2)$. Show that
$$\chi(G_1 \cup G_2) \leq \chi(G_1) \cdot \chi(G_2),$$
where, as usual, $\chi(G)$ denotes the minimum number of colours required to vertex-colour G. [H]
   Deduce that, if $G = (V, E)$ is a graph whose complement is planar, then $\chi(G) \geq |V|/5$. [H]

5. On the surface of a torus draw a cubic 'map' which cannot be coloured in six colours. [A]

15

Designs and codes

This chapter deals with one aspect of the 'design of experiments' where, loosely speaking, some different varieties of a commodity are to be compared as fairly as possible.

Example *Nine types of coffee are to be given to different households to compare. It is impracticable for any one household to compare all nine (after about the fourth sample they all begin to taste the same!). So twelve households are each to be given three brands to compare. Design the experiment so that each pair of the nine brands is compared in exactly one household.*

Solution Let the varieties of coffee be numbered 1–9 and the twelve households be given the following sets of three varieties to compare:

$\{1, 2, 3\}$, $\{4, 5, 6\}$, $\{7, 8, 9\}$, $\{1, 4, 7\}$, $\{1, 5, 9\}$, $\{2, 5, 8\}$,

$\{3, 6, 9\}$, $\{2, 6, 7\}$, $\{3, 4, 8\}$, $\{1, 6, 8\}$, $\{2, 4, 9\}$, $\{3, 5, 7\}$.

Then each pair from 1–9 is in precisely one of these sets. Note also (although it was not a stated requirement) that each variety will be tested in four households. □

That very simple example motivates our definition: a *design* (or, more fully, a *balanced incomplete block design*) consists of some proper subsets (or *blocks*) of a set of elements (or *varieties*), with each subset containing the same number (>1) of varieties and with the key property that each pair of varieties is in the same number of blocks. As we now see in our first theorem, it is also a consequence of that definition that each variety is then in the same number of blocks:

Theorem *Assume that we have a design with v varieties, b blocks each consisting of k ($1 < k < v$) varieties, and with each pair of varieties in precisely λ blocks. Then each of the varieties is in precisely r blocks, where*

$$r = \frac{bk}{v} = \frac{\lambda(v-1)}{k-1}.$$

Proof Let x be any particular variety of the design and assume that x is in precisely

r_x of the blocks. We consider all the blocks of the design and count, in two different ways, the grand total of all the pairs involving x.

Since x occurs in r_x blocks and in each of those blocks x can be paired with $k - 1$ other varieties, it follows that this grand total of pairs involving x is $(k - 1)r_x$. But altogether x can be paired with any one of the $v - 1$ other varieties and, by the definition of the design, each pair is in precisely λ blocks. So the grand total of pairs involving x is $\lambda(v - 1)$. Hence

$$(k - 1)r_x = \lambda(v - 1)$$

and

$$r_x = \frac{\lambda(v - 1)}{k - 1}.$$

In particular, r_x does not depend upon x and so each variety is in the same number of blocks, r say.

To establish the other part of the given relationship we now simply count, in two different ways, the total number of entries in all the blocks. This total is vr (each of the v varieties occurring r times) and it is also bk (each of the b blocks containing k varieties). Hence we have $vr = bk$ and

$$r = \frac{bk}{v}$$

as required. □

Therefore with this type of design there are associated five parameters:

v: the number of varieties,
b: the number of blocks,
r: the number of blocks containing any particular variety,
k: the number of varieties in each block,
λ: the number of blocks containing any particular pair of varieties,

and we shall refer to it as a '(v, b, r, k, λ) design'. So the solution of the coffee example above is a (9, 12, 4, 3, 1) design. Various notations are used in the study of designs and since, as the previous theorem shows, any three of the five parameters determine the other two, some authors refer to a design by stating only three of its parameters.

It is worth noting that if we are given five positive integers satisfying the relationships of the previous theorem (even if we also require that $b \geqslant v$, which turns out to hold in a design) then it does not necessarily follow that there exists a design with those parameters. For example $v = b = 43$, $r = k = 7$ and $\lambda = 1$ satisfy those relationships but, as we shall see in the exercises, no (43, 43, 7, 7, 1) design exists. In general designs are rather elusive to construct, although we shall see a couple of simple techniques later in the chapter and in the exercises.

As mentioned earlier, our particular designs are known more fully as 'balanced incomplete block designs' (or BIBDs). The study of the design of experiments actually encompasses much more than this specific sort of design; for example the 't-designs'

Designs and codes

require that each set of t varieties lies in a constant number of blocks (so that our designs are the '2-designs'). The interested reader can find fuller details in, for example, A.P. and D.J. Street's *Combinatorics of experimental design* listed in the bibliography.

Before proceeding to some further properties of designs we find a convenient way of representing them by means of matrices. Given a family of subsets $(A_1, \ldots, A_b)$ of a set $\{x_1, \ldots, x_v\}$ its *incidence matrix* M is the $b \times v$ matrix whose entry in the ith row and jth colour is given by

$$(i, j)\text{th entry of } M = \begin{cases} 1 & \text{if } x_j \in A_i, \\ 0 & \text{if } x_j \notin A_i. \end{cases}$$

So in this way we can consider the $b \times v$ incidence matrix of a family of sets formed by the blocks of a (v, b, r, k, λ) design: of course M depends upon the ordering of the blocks and of the varieties, but if M is an incidence matrix of a particular design then so is any matrix obtained by rearranging the rows and columns of M.

Example The $(9, 12, 4, 3, 1)$ design given in the previous example with varieties 1–9 and blocks

$\{1, 2, 3\}$, $\{4, 5, 6\}$, $\{7, 8, 9\}$, $\{1, 4, 7\}$, $\{1, 5, 9\}$, $\{2, 5, 8\}$,
$\{3, 6, 9\}$, $\{2, 6, 7\}$, $\{3, 4, 8\}$, $\{1, 6, 8\}$, $\{2, 4, 9\}$, $\{3, 5, 7\}$

has incidence matrix as shown, with each row representing a block in the obvious way:

$$M = \begin{pmatrix} 1 & 1 & 1 & 0 & 0 & 0 & 0 & 0 & 0 \\ 0 & 0 & 0 & 1 & 1 & 1 & 0 & 0 & 0 \\ 0 & 0 & 0 & 0 & 0 & 0 & 1 & 1 & 1 \\ 1 & 0 & 0 & 1 & 0 & 0 & 1 & 0 & 0 \\ 1 & 0 & 0 & 0 & 1 & 0 & 0 & 0 & 1 \\ 0 & 1 & 0 & 0 & 1 & 0 & 0 & 1 & 0 \\ 0 & 0 & 1 & 0 & 0 & 1 & 0 & 0 & 1 \\ 0 & 1 & 0 & 0 & 0 & 1 & 1 & 0 & 0 \\ 0 & 0 & 1 & 1 & 0 & 0 & 0 & 1 & 0 \\ 1 & 0 & 0 & 0 & 0 & 1 & 0 & 1 & 0 \\ 0 & 1 & 0 & 1 & 0 & 0 & 0 & 0 & 1 \\ 0 & 0 & 1 & 0 & 1 & 0 & 1 & 0 & 0 \end{pmatrix}$$

with b rows, v columns, k 1s in each row, and r 1s in each column.

Four of the parameters can be instantly seen within M as shown, but the parameter λ is more concealed. The fact that M is the incidence matrix of a (9, 12, 4, 3, 1) design means that any two varieties lie in precisely one of the blocks; i.e. any two columns of M have 1s together in precisely one place. In general any two columns of the matrix of a (v, b, r, k, λ) design will have 1s together in λ places. But a neater way of highlighting λ is to calculate the matrix product $M^{\mathsf{T}}M$, where ^{T} denotes the transpose: you can easily check that in this case it is

$$M^{\mathsf{T}}M = \begin{pmatrix} 4 & 1 & 1 & 1 & 1 & 1 & 1 & 1 & 1 \\ 1 & 4 & 1 & 1 & 1 & 1 & 1 & 1 & 1 \\ 1 & 1 & 4 & 1 & 1 & 1 & 1 & 1 & 1 \\ 1 & 1 & 1 & 4 & 1 & 1 & 1 & 1 & 1 \\ 1 & 1 & 1 & 1 & 4 & 1 & 1 & 1 & 1 \\ 1 & 1 & 1 & 1 & 1 & 4 & 1 & 1 & 1 \\ 1 & 1 & 1 & 1 & 1 & 1 & 4 & 1 & 1 \\ 1 & 1 & 1 & 1 & 1 & 1 & 1 & 4 & 1 \\ 1 & 1 & 1 & 1 & 1 & 1 & 1 & 1 & 4 \end{pmatrix}.$$

In general this product will have rs down the leading diagonal and λs elsewhere. □

Theorem *Let M be the incidence matrix of a family of b subsets of a set of v elements, with each subset containing precisely k elements $(1 < k < v)$. Then M is the incidence matrix of a (v, b, r, k, λ) design if and only if $M^{\mathsf{T}}M$ is the $v \times v$ matrix*

$$\begin{pmatrix} r & \lambda & \lambda & \ldots & \lambda \\ \lambda & r & \lambda & \ldots & \lambda \\ \lambda & \lambda & r & \ldots & \lambda \\ \vdots & \vdots & \vdots & & \vdots \\ \lambda & \lambda & \lambda & \ldots & r \end{pmatrix}.$$

Furthermore this $v \times v$ matrix then has positive determinant $(r + (v-1)\lambda)(r - \lambda)^{v-1}$.

Proof It is easy to check that the family of sets form a design if and only if $M^{\mathsf{T}}M$ has the required form. For, diagramatically, the (i, j)th entry of $M^{\mathsf{T}}M$ is given by

$$\boxed{i\text{th row of } M^{\mathsf{T}}} \cdot \underset{j\text{th column of } M}{\underline{}} = \underset{i\text{th column of } M}{\underline{}} \cdot \underset{j\text{th column of } M}{\underline{}} = \begin{cases} \text{number of 1s in the } i\text{th column of } M \ (i = j) \\ \text{number of places where there are} \\ \text{1s in both the } i\text{th and } j\text{th columns } (i \neq j) \end{cases}$$

and these will be constants r and λ, respectively, if and only if the sets form a (v, b, r, k, λ) design.

Elementary row and column operations on the $v \times v$ matrix enable its determinant to be calculated easily: adding all the rows to the first and then deducting the first column from each of the others gives

$$\det \begin{pmatrix} r & \lambda & \lambda & \lambda & \cdots & \lambda \\ \lambda & r & \lambda & \lambda & \cdots & \lambda \\ \lambda & \lambda & r & \lambda & \cdots & \lambda \\ \lambda & \lambda & \lambda & r & \cdots & \lambda \\ \vdots & \vdots & \vdots & \vdots & & \vdots \\ \lambda & \lambda & \lambda & \lambda & \cdots & r \end{pmatrix} = \det \begin{pmatrix} r+(v-1)\lambda & 0 & 0 & 0 & \cdots & 0 \\ \lambda & r-\lambda & 0 & 0 & \cdots & 0 \\ \lambda & 0 & r-\lambda & 0 & \cdots & 0 \\ \lambda & 0 & 0 & r-\lambda & \cdots & 0 \\ \vdots & \vdots & \vdots & \vdots & & \vdots \\ \lambda & 0 & 0 & 0 & \cdots & r-\lambda \end{pmatrix}$$

which is $(r + (v-1)\lambda)(r - \lambda)^{v-1}$.

Finally, to see that this determinant is positive it merely remains to prove that $r - \lambda > 0$. But this follows from the previous theorem since, as $1 < k < v$, we have

$$r = \frac{\lambda(v-1)}{k-1} > \lambda. \qquad \square$$

We are now ready to prove another connection between the parameters of a (v, b, r, k, λ) design, namely that $b \geqslant v$; in other words any experiment to compare v varieties fairly by means of a design needs at least that number of blocks. This inequality was first proved in 1940 by the distinguished statistician Sir Ronald Fisher.

Theorem (Fisher's inequality) *In a (v, b, r, k, λ) design $b \geqslant v$.*

Proof Assume that we have a (v, b, r, k, λ) design with $b < v$: we shall deduce a contradiction. Let M be an incidence matrix of the design: it is a $b \times v$ matrix and so it has fewer rows than columns. Add $v - b$ rows of zeros to M to form the $v \times v$ matrix N. Then, diagrammatically,

$$N = \begin{pmatrix} M \\ \hline 0 \end{pmatrix}$$

and

$$N^{\mathrm{T}} N = \begin{pmatrix} M \\ \hline 0 \end{pmatrix}^{\mathrm{T}} \begin{pmatrix} M \\ \hline 0 \end{pmatrix} = \begin{pmatrix} M^{\mathrm{T}} & 0 \end{pmatrix} \begin{pmatrix} M \\ \hline 0 \end{pmatrix} = M^{\mathrm{T}} M$$

which, as we saw in the previous theorem, has positive determinant.

Now N is a square matrix and so it has a well-defined determinant (which equals the determinant of N^T). Hence

$$(\det N)^2 = \det N^T \cdot \det N = \det(N^T N) = \det(M^T M) > 0.$$

But this contradicts the fact that N has a whole row of zeros (and hence has zero determinant). This contradiction shows that $b \geq v$ as required. □

So in any design there are at least as many blocks as varieties and, in some sense, the most economical designs are those for which the number of blocks equals the number of varieties. In such a design we also have $r = bk/v = k$ and so the parameters are (v, v, k, k, λ) and the design is called *symmetric*. Before proceeding to some properties of the symmetric designs we show how to construct a special type of them known as the 'cyclic' designs.

Example *Each of the following two matrices arises from a 'cyclic' construction where, after the first row is given, subsequent rows are obtained from the previous one by 'shifting' it one place to the right (and moving the last entry round to the front):*

$$\begin{pmatrix} 0 & 1 & 0 & 1 & 1 & 1 & 0 & 0 & 0 & 1 & 0 \\ 0 & 0 & 1 & 0 & 1 & 1 & 1 & 0 & 0 & 0 & 1 \\ 1 & 0 & 0 & 1 & 0 & 1 & 1 & 1 & 0 & 0 & 0 \\ 0 & 1 & 0 & 0 & 1 & 0 & 1 & 1 & 1 & 0 & 0 \\ 0 & 0 & 1 & 0 & 0 & 1 & 0 & 1 & 1 & 1 & 0 \\ 0 & 0 & 0 & 1 & 0 & 0 & 1 & 0 & 1 & 1 & 1 \\ 1 & 0 & 0 & 0 & 1 & 0 & 0 & 1 & 0 & 1 & 1 \\ 1 & 1 & 0 & 0 & 0 & 1 & 0 & 0 & 1 & 0 & 1 \\ 1 & 1 & 1 & 0 & 0 & 0 & 1 & 0 & 0 & 1 & 0 \\ 0 & 1 & 1 & 1 & 0 & 0 & 0 & 1 & 0 & 0 & 1 \\ 1 & 0 & 1 & 1 & 1 & 0 & 0 & 0 & 1 & 0 & 0 \end{pmatrix} \quad \begin{pmatrix} 0 & 1 & 0 & 0 & 1 & 1 & 0 & 1 & 0 & 1 & 0 \\ 0 & 0 & 1 & 0 & 0 & 1 & 1 & 0 & 1 & 0 & 1 \\ 1 & 0 & 0 & 1 & 0 & 0 & 1 & 1 & 0 & 1 & 0 \\ 0 & 1 & 0 & 0 & 1 & 0 & 0 & 1 & 1 & 0 & 1 \\ 1 & 0 & 1 & 0 & 0 & 1 & 0 & 0 & 1 & 1 & 0 \\ 0 & 1 & 0 & 1 & 0 & 0 & 1 & 0 & 0 & 1 & 1 \\ 1 & 0 & 1 & 0 & 1 & 0 & 0 & 1 & 0 & 0 & 1 \\ \mathbf{1} & \mathbf{1} & 0 & 1 & 0 & 1 & 0 & 0 & 1 & 0 & 0 \\ 0 & 1 & 1 & 0 & 1 & 0 & 1 & 0 & 0 & 1 & 0 \\ 0 & 0 & 1 & 1 & 0 & 1 & 0 & 1 & 0 & 0 & 1 \\ 1 & 0 & 0 & 1 & 1 & 0 & 1 & 0 & 1 & 0 & 0 \end{pmatrix}$$

<table>
<tr><td align="center">the incidence matrix of
an (11, 11, 5, 5, 2) design</td><td align="center">not the incidence matrix
of a design</td></tr>
</table>

If we consider the varieties here to be 0, 1, 2, ..., 10 then the first block is $\{1, 3, 4, 5, 9\}$ and each subsequent block is obtained by adding 1 (modulo 11) to each entry of the previous block. The process gives rise to a design in this case.

Here the same process, but with first block $\{1, 4, 5, 7, 9\}$, does not lead to a design. For example the pair of varieties $\{0, 1\}$ is in just one of the sets whereas the pair $\{6, 9\}$ is in three of the sets, as highlighted.

If we construct the table showing the differences (mod 11) of members of the first block $\{1, 3, 4, 5, 9\}$ we get

$-$(mod 11)	1	3	4	5	9
1	0	9	8	7	3
3	2	0	10	9	5
4	3	1	0	10	6
5	4	2	1	0	7
9	8	6	5	4	0

Notice that each non-zero entry occurs in the body of the table the same number of times. (In fact each appears twice and it is no coincidence that $\lambda = 2$ for this particular design.)

In this case the differences (mod 11) of members of the first block $\{1, 4, 5, 7, 9\}$ are

$-$(mod 11)	1	4	5	7	9
1	0	8	7	5	3
4	3	0	10	8	6
5	4	1	0	9	7
7	6	3	2	0	9
9	8	5	4	2	0

Notice that the answer 1 occurs in the body of this table just once whereas the number 3 occurs three times. In fact in both examples the number of blocks containing i and j equals the number of times $|i - j|$ occurs in the table of differences. □

A subset P of $\{0, 1, 2, \ldots, v-1\}$ is called a *perfect difference set* (mod v) if in P's table of difference modulo v each of $1, 2, \ldots, v-1$ occurs the same number of times. So in the above example $\{1, 3, 4, 5, 9\}$ is a perfect difference set (mod 11), and in fact forms the first set of a design constructed by that 'cyclic' process. However, $\{1, 4, 5, 7, 9\}$ is not a perfect difference set (mod 11), and for this set the process fails to produce a design.

Theorem *Let $B_0 = \{b_1, \ldots, b_k\}$ be a subset of $\{0, 1, 2, \ldots, v-1\}$ with $1 < k < v$, let $+$ denote addition modulo v, and for $0 \leq i \leq v-1$ let B_i be the set $B_i = \{b_1 + i, \ldots, b_k + i\}$. Then the sets $B_0, B_1, \ldots, B_{v-1}$ form a design if and only if B_0 is a perfect difference set (mod v). (Such a design is then called* cyclic.*)*

Proof Consider the sets $B_0, \ldots, B_{v-1}$: they all have the same size. So to see if they form a design we must check whether any two of the v varieties lie in the same number of sets. The cyclic nature of the sets means that, for example, varieties 0 and 3 will occur in exactly the same number of sets as 1 and 4 or as 2 and 5 etc. So we shall take any j in $\{1, 2, \ldots, v-1\}$ and consider the number of the sets $B_0, \ldots, B_{v-1}$ which contain both the varieties 0 and j (and it makes things a little easier if we assume, without loss in generality, that $0 \notin B_0$).

When will 0 and j both appear in the set B_i? It is easy to see that it can only happen in the following way:

$$B_i = \{b_1 + i, \ldots, \underbrace{b_\alpha + i}_{= 0?}, \ldots, \underbrace{b_\beta + i}_{= j?}, \ldots, b_k + i\}.$$

$$\begin{array}{cc} \text{if and only if} & \text{if and only if} \\ i = v - b_\alpha & b_\beta + i \equiv j \pmod{v} \end{array}$$

We claim that 0 and j are both in the set B_i, as shown, if and only if $i = v - b_\alpha$ and $b_\beta - b_\alpha \equiv j \pmod{v}$. For if both 0 and j are in B_i as shown then

$$i = v - b_\alpha \quad \text{and} \quad j \equiv b_\beta + i = b_\beta + (v - b_\alpha) \equiv b_\beta - b_\alpha \pmod{v}$$

as claimed. Conversely, if $i = v - b_\alpha$ and $b_\beta - b_\alpha \equiv j \pmod{v}$, then

$$i = v - b_\alpha \quad \text{and} \quad b_\beta + i = (b_\beta - b_\alpha) + (b_\alpha + i) \equiv j \pmod{v},$$

so that both 0 and j are in the set B_i. Hence each time that $j \equiv b_\beta - b_\alpha \pmod{v}$ it follows that 0 and j are both in the set B_{v-b_α}, and all such occurrences happen this way. Therefore there is a one-to-one correspondence between the times that 0 and j are together in one of the sets $B_0, \ldots, B_{v-1}$ and the times that two members of B_0 differ by j modulo v (which is the number of times j appears in B_0's table of differences modulo v).

Therefore any two varieties will appear in the same number of sets if and only if each number j ($1 \leq j < v$) occurs the same number of times in B_0's table of differences modulo v; i.e. if and only if B_0 is a perfect difference set (mod v). □

For that theorem to be any practical use in the construction of designs we need to know a systematic way of finding perfect difference sets.

Example *Form the powers of 2 modulo 19 to give*:

$$2^1 = 2, \quad 2^2 = 4, \quad 2^3 = 8, \quad 2^4 = 16, \quad 2^5 = (32 \equiv)13, \quad 2^6 = 7, \quad 2^7 = 14,$$
$$2^8 = 9, \quad 2^9 = 18, \quad 2^{10} = 17, \quad 2^{11} = 15, \quad 2^{12} = 11, \quad 2^{13} = 3, \quad 2^{14} = 6,$$
$$2^{15} = 12, \quad 2^{16} = 5, \quad 2^{17} = 10, \quad 2^{18} = 1, \ldots$$

Then look at the set of even powers, namely $Q = \{1, 4, 5, 6, 7, 9, 11, 16, 17\}$. As the following difference table shows, this is a perfect difference set (mod 19):

$-$ (mod 19)	1	4	5	6	7	9	11	16	17
1	0	16	15	14	13	11	9	4	3
4	3	0	18	17	16	14	12	7	6
5	4	1	0	18	17	15	13	8	7
6	5	2	1	0	18	16	14	9	8
7	6	3	2	1	0	17	15	10	9
9	8	5	4	3	2	0	17	12	11
11	10	7	6	5	4	2	0	14	13
16	15	12	11	10	9	7	5	0	18
17	16	13	12	11	10	8	6	1	0

each non-zero number occurs four times

Hence starting with the set $Q = \{1, 4, 5, 6, 7, 9, 11, 16, 17\}$ and successively adding 1 modulo 19 will lead to a cyclic $(19, 19, 9, 9, 4)$ design. □

In that example the numbers 1, 4, 5, 6, 7, 9, 11, 16, 17 are called the 'quadratic residues' under multiplication modulo 19 because they turn out to be the perfect squares under that operation. We now see how the process in that example generalises to any $v = 4n - 1$ which is prime: in the theorem and its proof we assume some of the algebraic properties of multiplication modulo v.

Theorem *Let v be a prime of the form $4n - 1$ for some integer n. Let θ generate $\{1, \ldots, v-1\}$ under multiplication modulo v; i.e. $\{\theta, \theta^2, \ldots, \theta^{v-1}\}$ is the whole of $\{1, \ldots, v-1\}$. Then $\{\theta^2, \theta^4, \ldots, \theta^{v-1}\}$, the set of 'quadratic residues (mod v)', is a perfect difference set (mod v) and forms the first set of a cyclic $(4n - 1, 4n - 1, 2n - 1, 2n - 1, n - 1)$ design.*

Proof Throughout this proof we shall be working in the set $\{0, 1, \ldots, v-1\}$ under the usual arithmetic operations modulo v (making it a field, and in particular there is a '-1' behaving in a natural way). With that understanding, $\theta^{v-1} = 1$, $\theta^v = \theta$, $\theta^{v+1} = \theta^2$, etc., so that no even power of θ is ever equal to an odd power, and Q consists precisely of all the even powers of θ. Note also that, since $(-1)^2 = 1 = \theta^{v-1}$, it follows that $-1 = \theta^{(v-1)/2} = \theta^{2n-1}$.

Now let $1 \leq i, j < v$ and let $j \cdot i^{-1} = \theta^k$. We shall show how to each entry of i in Q's difference table (modulo v) there corresponds a unique entry of j. So look at any particular entry of i in that table, $i = \theta^{2a} - \theta^{2b}$ say. Then

$$j = \theta^k \cdot i = \theta^k(\theta^{2a} - \theta^{2b}) = \theta^{2a+k} - \theta^{2b+k}.$$

If k is even, then this already expresses j as a difference of members of Q. If, on the other hand, k is odd then

$$j = \theta^{2a+k} - \theta^{2b+k} = (-1) \cdot (\theta^{2b+k} - \theta^{2a+k}) = \theta^{2n-1} \cdot (\theta^{2b+k} - \theta^{2a+k})$$
$$= \theta^{2b+k+2n-1} - \theta^{2a+k+2n-1}$$

and this again expresses j as a difference of two members of Q.

So to each entry of i in Q's difference table there corresponds an entry of j. A little thought will show you that different entries of i will lead, by the above process, to different entries of j. Hence j occurs in that table at least as many times as i and, by symmetry, it follows that i and j occur the same number of times. Thus any two members of $\{1, \ldots, v-1\}$ occur the same number of times in Q's difference table, and Q is a perfect difference set as claimed.

Therefore, by the previous theorem, Q is the first set of a cyclic design: it has parameters
$$b = v = 4n - 1, \qquad r = k = |Q| = \tfrac{1}{2}(v - 1) = 2n - 1$$
and

$$\lambda = \frac{r(k-1)}{v-1} = \frac{(2n-1)(2n-2)}{4n-2} = n-1,$$

as required. □

When, for prime n, we constructed $n-1$ mutually orthogonal Latin squares (page 63), we used modular arithmetic and noted that, by using the Galois field $GF(n)$, the process generalised to the cases when n was a power of a prime. This was because the crucial part of the construction merely depended upon the existence of a field on $\{0, 1, \ldots, n-1\}$. By a similar argument the previous two theorems can be generalised to construct cyclic $(4n-1, 4n-1, 2n-1, 2n-1, n-1)$ designs whenever $v = 4n-1$ is a power of a prime. Details of this generalisation, and of other constructions of cyclic designs, can be found in the book by Street and Street mentioned earlier.

The so-called 'symmetric' designs (i.e. those with $b = v$) have earned that title because of a special property which we now illustrate.

Example Let us look at the symmetric $(7, 7, 3, 3, 1)$ design with blocks $\{1, 2, 3\}$, $\{2, 4, 6\}$, $\{3, 5, 6\}$, $\{1, 4, 5\}$, $\{1, 6, 7\}$, $\{2, 5, 7\}$, $\{3, 4, 7\}$. It has incidence matrix

$$M = \begin{pmatrix} 1 & 1 & 1 & 0 & 0 & 0 & 0 \\ 0 & 1 & 0 & 1 & 0 & 1 & 0 \\ 0 & 0 & 1 & 0 & 1 & 1 & 0 \\ 1 & 0 & 0 & 1 & 1 & 0 & 0 \\ 1 & 0 & 0 & 0 & 0 & 1 & 1 \\ 0 & 1 & 0 & 0 & 1 & 0 & 1 \\ 0 & 0 & 1 & 1 & 0 & 0 & 1 \end{pmatrix}$$

Note in this case that the transpose of M is given by

$$M^T = \begin{pmatrix} 1 & 0 & 0 & 1 & 1 & 0 & 0 \\ 1 & 1 & 0 & 0 & 0 & 1 & 0 \\ 1 & 0 & 1 & 0 & 0 & 0 & 1 \\ 0 & 1 & 0 & 1 & 0 & 0 & 1 \\ 0 & 0 & 1 & 1 & 0 & 1 & 0 \\ 0 & 1 & 1 & 0 & 1 & 0 & 0 \\ 0 & 0 & 0 & 0 & 1 & 1 & 1 \end{pmatrix}$$

which itself is the matrix of another $(7, 7, 3, 3, 1)$ design with blocks $\{1, 4, 5\}$, $\{1, 2, 6\}$, $\{1, 3, 7\}$, $\{2, 4, 7\}$, $\{3, 4, 6\}$, $\{2, 3, 5\}$, $\{5, 6, 7\}$: this new design is called a 'dual' of the original one. □

Designs and codes

Theorem *If M is an incidence matrix of a symmetric design, then so is the matrix M^T (and the new design is called a* dual *of the original one).*

Proof Let M be the incidence matrix of (v, v, k, k, λ) design. Then M is a $v \times v$ matrix of 0s and 1s with precisely k 1s in each row and in each column. Also, if we let Λ be the $v \times v$ matrix all of whose entries are λ, and let I be the $v \times v$ identity matrix, then by the theorem on page 182

$$M^T M = \begin{pmatrix} k & \lambda & \lambda & \cdots & \lambda \\ \lambda & k & \lambda & \cdots & \lambda \\ \lambda & \lambda & k & \cdots & \lambda \\ \vdots & \vdots & \vdots & & \vdots \\ \lambda & \lambda & \lambda & \cdots & k \end{pmatrix} = \Lambda + (k - \lambda)I.$$

Now M^T is a $v \times v$ matrix with precisely k 1s in each row and so, by that same theorem (applied to M^T rather than M), to show that M^T is the incidence matrix of a (v, v, k, k, λ) design we must show that $(M^T)^T M^T$ is also equal to $\Lambda + (k - 1)I$; i.e. we must show that MM^T equals $M^T M$ (this *does* require proof: it is not in general true that the multiplication of a matrix with its transpose is commutative).

Note that, as before, $\det(M^T M) > 0$ and as M is a square matrix in this case we have

$$(\det M)^2 = \det M^T \cdot \det M = \det(M^T M) > 0.$$

It follows that $\det M \neq 0$ and that M has an inverse. Note also that, as each row and each column of M contains exactly k 1s, it follows that

$$M \begin{pmatrix} \lambda & \lambda & \lambda & \cdots & \lambda \\ \lambda & \lambda & \lambda & \cdots & \lambda \\ \lambda & \lambda & \lambda & \cdots & \lambda \\ \vdots & \vdots & \vdots & & \vdots \\ \lambda & \lambda & \lambda & \cdots & \lambda \end{pmatrix} = \begin{pmatrix} k\lambda & k\lambda & k\lambda & \cdots & k\lambda \\ k\lambda & k\lambda & k\lambda & \cdots & k\lambda \\ k\lambda & k\lambda & k\lambda & \cdots & k\lambda \\ \vdots & \vdots & \vdots & & \vdots \\ k\lambda & k\lambda & k\lambda & \cdots & k\lambda \end{pmatrix} = \begin{pmatrix} \lambda & \lambda & \lambda & \cdots & \lambda \\ \lambda & \lambda & \lambda & \cdots & \lambda \\ \lambda & \lambda & \lambda & \cdots & \lambda \\ \vdots & \vdots & \vdots & & \vdots \\ \lambda & \lambda & \lambda & \cdots & \lambda \end{pmatrix} M$$

i.e. $M\Lambda = \Lambda M$. Hence

$$\begin{aligned} MM^T &= (MM^T)(MM^{-1}) \\ &= M(M^T M)M^{-1} \\ &= M(\Lambda + (k-\lambda)I)M^{-1} \\ &= M\Lambda M^{-1} + (k-\lambda)MIM^{-1} \\ &= \Lambda MM^{-1} + (k-\lambda)I \\ &= \Lambda + (k-\lambda)I \\ &= M^T M \end{aligned} \right\} \text{ since } M\Lambda = \Lambda M$$

as claimed.

Therefore M^T is a $v \times v$ matrix with k 1s in each row and with

$$(M^T)^T M^T = \begin{pmatrix} k & \lambda & \lambda & \cdots & \lambda \\ \lambda & k & \lambda & \cdots & \lambda \\ \lambda & \lambda & k & \cdots & \lambda \\ \vdots & \vdots & \vdots & & \vdots \\ \lambda & \lambda & \lambda & \cdots & k \end{pmatrix}$$

Hence, as commented above, the theorem on page 182 shows that M^T is the incidence matrix of a (v, v, k, k, λ) design, as required. □

First corollary *Any two blocks of a (v, v, k, k, λ) design intersect in precisely λ points.*

Proof Let M be an incidence matrix of the (v, v, k, k, λ) design. Then, by the theorem, M^T is also an incidence matrix of some (v, v, k, k, λ) design. Hence any two columns of M^T have 1s together in precisely λ places; i.e. any two rows of M have 1s together in precisely λ places.

Now to consider the intersection of two of the blocks of the design look at the corresponding two rows of M. Then the number of elements in the intersection of the two blocks equals the number of places in which those two rows both have entries of 1s: by the above comments this equals λ. □

Second corollary *Any two rows of an incidence matrix of a (v, v, k, k, λ) design differ in precisely $2(k - \lambda)$ places.*

Proof It is clear from the first corollary that any two rows of an incidence matrix of a (v, v, k, k, λ) design (with the columns re-ordered for convenience) look like

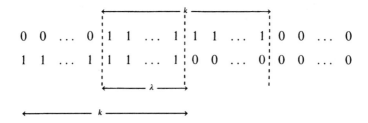

Hence there are $k - \lambda$ places with 1s in the first row and 0s in the second, and a further $k - \lambda$ places with 0s in the first row and 1s in the second, and the corollary follows. □

Hence the symmetric designs, apart from being the most 'economical', have the

Designs and codes

special property of duality and, in particular, any two of their blocks meet in the same number of points. These properties can be observed in the cyclic designs constructed above, but this is not the only class of symmetric designs which we have met. Although you may not recognise the fact, we have seen another type of symmetric design in our work on finite projective planes (page 65): it turns out that the modern study of designs incorporates that earlier work.

Example *Here again is a finite projective plane of order 2: its lines are $\{1, 2, 3\}, \{2, 4, 6\}, \{3, 5, 6\}, \{1, 4, 5\}, \{1, 6, 7\}, \{2, 5, 7\}$ and $\{3, 4, 7\}$. Note that those lines form the blocks of a $(7, 7, 3, 3, 1)$ design. (Also, the previous theorem about dual designs is a generalisation of the idea of the duality of the points and lines in this geometry which we referred to in chapter 5.)*

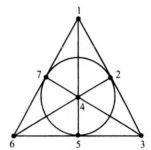

Theorem *A finite projective plane of order n is equivalent to an $(n^2 + n + 1, n^2 + n + 1, n + 1, n + 1, 1)$ design; i.e. the lines of a finite projective plane of order n form the blocks of such a design and vice versa.*

Proof Assume first that we are given a finite projective plane of order n. Recall that this consists of a set of points and a collection of 'lines' (which are subsets of those points) with the properties that
(i) any two points lie in precisely one line,
(i)' any two lines meet in precisely one point,
(ii) each line contains $n + 1$ points,
(ii)' each point lies in $n + 1$ lines.
Do these lines form the blocks of a design? Of course they do because firstly they are the same size (property (ii)) and secondly any two points lie in the same number of blocks/lines (property (i)). Hence we certainly have a design with $k = n + 1$ and with $\lambda = 1$. Also property (i)' tells us that $r = n + 1$. Hence the remaining two parameters b and v are given by

$$n + 1 = \frac{b(n + 1)}{v} = \frac{v - 1}{n}$$

from which it follows that $b = v = n^2 + n + 1$. Therefore we do have an $(n^2 + n + 1, n^2 + n + 1, n + 1, n + 1, 1)$ design as claimed. (In fact we used very similar arguments in solving exercise 7 on page 70 before we had even met designs.)

Conversely, assume that we are given an $(n^2 + n + 1, n^2 + n + 1, n + 1, n + 1, 1)$ design. It is immediate that, if we regard the blocks as lines, then the parameters imply properties (i), (ii) and (ii)':

$$(n^2 + n + 1, \quad n^2 + n + 1, \quad n + 1, \quad n + 1, \quad 1)$$
$$\Downarrow \qquad\qquad \Downarrow \qquad \Downarrow$$
$$\text{(ii)}' \qquad\qquad \text{(ii)} \qquad \text{(i)}$$

Finally, the remaining property (i)' is immediate from the first corollary above which tells us that any two blocks of this design intersect in exactly one point. He..ce the blocks of the design do form the lines of a finite projective planes of order n, and the theorem is proved. □

So our method of constructing finite projective planes by using Latin squares (pages 66–69) gives us another method of constructing some symmetric designs. Also the non-existence of a finite projective plane of order 6 discussed there is equivalent to the non-existence of a (43, 43, 7, 7, 1) design, and we derive this in the exercises. Similarly a finite projective plane of order 10 is equivalent to a (111, 111, 11, 11, 1) design, and it was in this form that the non-existence was established by a computer search in 1989.

It turns out that the symmetric designs are useful not only in setting up fair experiments but also in another modern application of mathematics to 'error-correcting codes'. Therefore we now discuss the general theory of such codes before using designs to construct some. Suppose that you wish to send a message electronically: then basically that message must be translated into an electronic language which, for our purposes, will mean coding it into a string of 0s and 1s. (These are sometimes known as the 'binary' codes.)

Example *As a very simple example imagine that you will only ever want to send one of four words 'North', 'South', 'East' or 'West'. The simplest binary code is obtained by translating each of those words into the following 'codewords':*

$$N \to 00 \quad S \to 01 \quad E \to 10 \quad W \to 11$$

(*In our codes each codeword will use the same number of digits, that number being called the* length *of the code. So in this example the code has length 2.*) □

Because of the possibility of errors in sending electronic messages it is sometimes desirable to make a code a little more foolproof by building in some checking facility.

Example revisited *In the simple code of the previous example a single error can cause the completely wrong message to be received. For example an error in just one of the digits when transmitting 'North' could result in it being read as 'South'. The following code of length 3 is a refinement of that earlier one:*

$$N \to 000 \quad S \to 011 \quad E \to 101 \quad W \to 110$$

Designs and codes

The extra 'check' digit has been chosen so that each of the codewords contains an even number of 1s. So, if an error occurs in a single digit during the transmission of one of those words, then the message received will contain an odd number of 1s. The received message will therefore not be one of the acceptable codewords and the receiver will know that an error has occurred: the transmitter of the message can then be asked to repeat it. This is therefore an example of a code which can recognise an error. (A similar principle, in a much more complicated code, applies when the supermarket cashier electronically scans the bar-code on groceries: if there is an error then the computerised till can recognise that fact and it 'beeps' to tell the cashier to try again.) ☐

That example shows how a code can detect errors (or at least a limited number of them). Sometimes, however, it is impracticable to tell the transmitter to repeat a message (for example in the case of the coded weather data from an orbiting satellite or the coded sounds from a compact disc) and in these cases we need the code to be refined even further.

Example re-revisited *If, using the previous code, the message 001 arrived, then (as this is not one of the acceptable codewords) the receiver would know that an error had occurred. But, even assuming that there was an error in just one digit, it is impossible to tell whether the message intended was 000 (for 'North') or 011 (for 'South') or 101 (for 'East'). Hence (if possible) the receiver would ask for the message for be repeated. However we can refine the code to the following:*

$$N \to 000111 \quad S \to 011010 \quad E \to 101100 \quad W \to 110001$$

Although the code is twice the length, it has great advantages over the previous one. Firstly it can detect that an error has occurred if up to three digits are incorrectly transmitted in a word. That is because it is impossible by three or fewer changes to make one of the acceptable codewords into another. But the major improvement is that (assuming our equipment is reasonably reliable) in this case the receiver can actually correct *the message without having to ask for it to be repeated. For example, suppose that the received message is 111001: this is not an acceptable codeword and so an error must have occurred. What was the intended message?*

'North'? i.e. 000111: this would need five changes to make it into 111001.
'South'? i.e. 011010: this would need three changes to make it into 111001.
'East'? i.e. 101100: this would need three changes to make it into 111011.
'West'? i.e. 110001: this would need one change to make it into 111001.

On the assumption that the equipment is so reliable that more than one error is extremely unlikely, the receiver will decode the message as 'West', that being the word which requires fewest changes to make the received message. ☐

That is a simple example of an 'error-correcting code': it can detect errors if fewer than four errors are made per word, and it can actually correct them if fewer than two errors are made per word. The crucial property of the codewords which gives them this error-correcting capability is that it takes four changes of the digits to make any one codeword into another. We now state the simple generalisation of that which gives the key to the theory of error-correcting codes.

Theorem *Let the codewords of a code have the property that it takes at least d changes of its digits to make any one codeword into another. Then the code can*
(i) *detect errors if fewer than d are made in each word;*
(ii) *correct errors if fewer than $\frac{1}{2}d$ are made in each word.*

Proof (i) It takes at least d changes to make one codeword into another. So, if at least one but fewer than d changes occur in the transmission of a codeword, then the received string of 0s and 1s cannot be an acceptable codeword. Hence the receiver will realise in this case that an error has occurred.

(ii) In this case the maximum number of errors is even lower and so once again errors will be detected. But if there are less than $\frac{1}{2}d$ changes between the intended codeword and the actual received string of 0s and 1s then how many changes are there between that received string and any *other* codeword? Consider any such codeword and let there be c changes as shown:

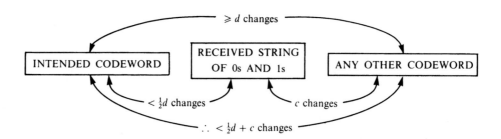

That diagram of changes shows that $\frac{1}{2}d + c > d$ and hence that $c > \frac{1}{2}d$. Therefore the intended codeword is the only one of the codewords which can be obtained from the received string by fewer than $\frac{1}{2}d$ changes. So in this case the intended codeword is uniquely determined as the codeword requiring fewest changes from the received string, and the receiver can work out the intended message despite the errors. □

The general theory of codes can be found, for example, in R. Hill's *A first course in coding theory* listed in the bibliography. But our interest is simply in those codes which can be derived from the symmetric designs. We saw in the second corollary on page 190 that any two rows of the incidence matrix of a (v, v, k, k, λ) design differ in precisely $2(k - \lambda)$ places. The following result is therefore immediate:

Designs and codes

Theorem *If the codewords of a code consist of the rows of the incidence matrix of a (v, v, k, k, λ) design then the code can*
(i) detect errors if fewer than $2(k - \lambda)$ are made in each word;
(ii) correct errors if fewer than $k - \lambda$ are made in each word. □

The codes derived from designs in that way have two special properties: firstly each of the words has the same 'weight' (i.e. contains the same number of 1s) and secondly any two of the codewords differ in exactly the same number of places. For codes with these special properties the ones derived from designs are the best possible for, as we shall see in the exercises, any such code of length v can have at most v codewords. So a $b \times v$ matrix whose rows are the codewords of such a code must have $b \leq v$ and, by Fisher's inequality, a $b \times v$ incidence matrix of a design must have $b \geq v$. As we have just seen the matrices of the designs with the minimum number of blocks turn out to be matrices of these special codes with the maximum of codewords, and the two concepts have come together in another 'minimax' theorem. The reader who wishes to pursue the many fascinating interrelationships between designs and codes can consult the advanced text *Graph theory, coding theory and block designs* by P.J. Cameron and J.H. Van Lint.

Codes of length v obtained from designs as above have v codewords: we conclude this chapter with an example illustrating how in some cases it is possible to then derive a code of length $v + 1$ with $2(v + 1)$ codewords and yet with the same error-correcting capabilities.

Example *The matrix M is an incidence matrix of a $(7, 7, 3, 3, 1)$ design:*

$$M = \begin{pmatrix} 1 & 1 & 1 & 0 & 0 & 0 & 0 \\ 0 & 1 & 0 & 1 & 0 & 1 & 0 \\ 0 & 0 & 1 & 0 & 1 & 1 & 0 \\ 1 & 0 & 0 & 1 & 1 & 0 & 0 \\ 1 & 0 & 0 & 0 & 0 & 1 & 1 \\ 0 & 1 & 0 & 0 & 1 & 0 & 1 \\ 0 & 0 & 1 & 1 & 0 & 0 & 1 \end{pmatrix}$$

As expected, any two rows of M differ in precisely four places. So if the rows of M are used as the codewords of a code then it can detect errors if less than four are made in a word, and it can correct errors if less than two are made.

Now let M' be the 7×7 matrix obtained from M by replacing 1s by 0s and 0s by 1s, and let N be the 16×8 matrix constructed as follows:

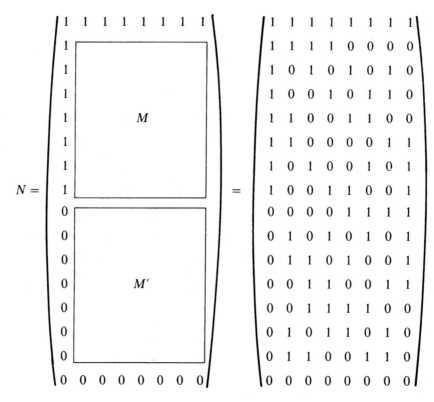

Then, as can be readily checked, any two rows of N differ in four or more places. So if the rows of N are used as the codewords of a code then it has the same error-correcting capabilities as before, but by increasing the length by 1 we have been able to more than double the number of codewords. □

In the exercises we shall see that the only symmetric designs capable of producing extended codes as in that example are the $(4n - 1, 4n - 1, 2n - 1, 2n - 1, n - 1)$ designs, like the cyclic ones we constructed earlier.

Exercises

1. Given integers k and v with $1 < k < v$ show that there exists a
$$\left(v, \binom{v}{k}, \binom{v-1}{k-1}, k, \binom{v-2}{k-2}\right) \text{ design.}$$

2. Given a (v, b, r, k, λ) design show that there exists a $(v, b, b - r, v - k, b + \lambda - 2r)$ design. [H]

3. Given a (v, b, r, k, λ) design and a particular block B of it, for $0 \leqslant i \leqslant k$ let x_i denote the number of other blocks which intersect B in precisely i varieties. Show that

Designs and codes

(i) $\sum_{i=0}^{k} x_i = b - 1;$

(ii) $\sum_{i=0}^{k} i x_i = k(r - 1);$ [H]

(iii) $\sum_{i=0}^{k} i(i - 1) x_i = k(k - 1)(\lambda - 1).$ [H]

4. Show that if a (v, v, k, k, λ) design exists then $k + \lambda(v - 1)$ is a perfect square. Deduce that if v is even then $k - \lambda$ is also a perfect square. [H]

5. Show that if n is either a prime or a power of a prime then there exists an $(n^2, n^2 + n, n + 1, n, 1)$ design. [H]

6. A mathematician claims to be able to use trees to construct designs. For an integer $n > 2$ he takes a complete graph K_n (with vertex-set V and edge-set E) and he then uses the edges as the varieties of a design. For the blocks he takes each subset of E which forms a tree on the vertex-set V.
e.g. For $n = 3$:

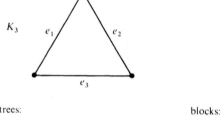

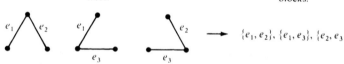

(giving a $(3, 3, 2, 2, 1)$ design). Show that if $n > 3$ this process fails to produce a design. [H]

7. (*Kirkman's schoolgirl problem*, 1850) Is it possible for 15 schoolgirls to walk in five rows of three on each day of the week so that in the course of the week each girl has been in the same row with each of the other girls exactly once? [A]

8. A *Steiner triple system* is a collection of b triple-element subsets of a set of v elements so that any pair of the v elements lies in exactly one of the triples. Show that in such a design
   (i) v is odd,
   (ii) $b = \frac{1}{6}v(v - 1),$
   (iii) v has the form $6n + 1$ or $6n + 3$ for some integer $n \geq 0$.
   (In fact Steiner triple systems are known to exist for all v satisfying (iii).)

9. Show that, if $B \subseteq \{0, 1, \ldots, v - 1\}$ is a perfect difference set (mod v) with $|B| = k$, then $k(k - 1)$ is divisible by $v - 1$.

In the theorem on page 187 we constructed such a set with $k = \frac{1}{2}(v - 1)$. Show that if such a set exists then $v = 4n - 1$ for some integer n.

10. (For keener readers to show that no (43, 43, 7, 7, 1) design exists – or equivalently that there is no finite projective plane of order 6.)

(i) Let Q be a square matrix. Show that there exists a diagonal matrix D all of whose diagonal entries are ± 1 so that the matrix $Q + D$ has non-zero determinant. [H]

(ii) Let P be a 44×44 matrix with rational entries. Show that there exist rational numbers $x_1, \ldots, x_{43}$ with

$$P \begin{pmatrix} x_1 \\ \vdots \\ x_{43} \\ 1 \end{pmatrix} = \begin{pmatrix} y_1 \\ \vdots \\ y_{43} \\ y \end{pmatrix}$$

where $y_1 = \pm x_1, \ldots, y_{43} = \pm x_{43}$. [H]

(iii) By considering the 4×4 matrix

$$K = \begin{pmatrix} 2 & 1 & 1 & 0 \\ 1 & -2 & 0 & -1 \\ 1 & 0 & -2 & 1 \\ 0 & 1 & -1 & -2 \end{pmatrix}$$

which has $K^T K = 6I_4$, find a 44×44 matrix L with $L^T L = 6I_{44}$ (where I_n denotes the $n \times n$ identity matrix). [A]

(iv) Show that there exist no rational numbers x, y with $6y^2 = x^2 + 1$. [H]

(v) We are now ready to show that no (43, 43, 7, 7, 1) design exists. Assume that such a design does exist, let M be its incidence matrix, let N be the 44×44 matrix

$$N = \begin{pmatrix} \boxed{M} & \begin{matrix} 0 \\ 0 \\ \vdots \\ 0 \end{matrix} \\ 0 \; 0 \; \cdots \; 0 & 1 \end{pmatrix}$$

and let $P = \frac{1}{6} L N^T$, where L is as in part (iii) above. Then apply (ii) to P to give rational numbers $x_1, \ldots, x_{43}$ and y as stated there and let $x = x_1 + \cdots + x_{43}$. Show that this x and y satisfy $6y^2 = x^2 + 1$ and deduce the required contradiction. [H]

Designs and codes

11. Assume that, with our equipment, in the transmission of a binary code there is an independent probability p that an error will occur in any particular digit. In the N/S/E/W examples in the chapter we met two error-correcting codes. The first was of length 3 and could detect up to one error in a codeword, and the second was of length 6 and could detect up to three errors in a codeword. Calculate, in terms of p,

   (i) the probability of more than one error in a codeword transmitted using the first code; [A]

   (ii) the probability of more than three errors in a codeword transmitted using the second code. [A]

   Calculate these probabilities in the cases $p = 0.75$, $p = 0.3$ and $p = 0.1$. You will see that for small p the second code is far more reliable. [A]

12. Consider a collection of v subsets of a set of b elements such that each of the subsets contains r elements and the intersection of any two of the sets contains λ elements. By counting the grand total of occurrences of elements in pairs of sets, or otherwise, show that

$$b((v-1)\lambda + r) \geqslant vr^2.$$

Show also that equality holds if and only if each of the b elements is in the same number of subsets, k say, where

$$r = \frac{bk}{v} = \frac{\lambda(v-1)}{k-1}. \quad [\text{H}]$$

13. Assume that we have a code of length v consisting of b codewords each of the same weight (i.e. each containing the same number of 1s) and such that any two of the codewords differ in the same number of places. Show that $b \leqslant v$. [H]

14. Let H be an $m \times m$ matrix with each entry ± 1, with its first row and column consisting entirely of 1s and with $HH^T = mI_m$, where $m > 2$ and I_m is the $m \times m$ identity matrix. (Such an H is known as a *normalised Hadamard matrix*.) Show that

   (i) each row and each column of H (apart from the first) contains exactly $m/2$ 1s; [H]

   (ii) given any two columns of H (apart from the first) they have 1s together in precisely $m/4$ places. [H]

   Now let $m = 4n$ and let M be the matrix obtained from H by deleting its first row and column and replacing all its -1s by 0s. Show that M is an incidence matrix of a $(4n-1, 4n-1, 2n-1, 2n-1, n-1)$ design.

15. Let M be an incidence matrix of a (v, v, k, k, λ) design. Then, as we have seen, any two rows of M differ in $2(k - \lambda)$ places and they can be used as the codewords of an error-correcting code. Now, as in the last example of the chapter, let M' be the matrix obtained from M by replacing 1s by 0s and 0s by 1s, and let N be the $2(v+1) \times (v+1)$ matrix of the form

$$N = \begin{pmatrix} 1 & 1 & 1 & \cdots & 1 \\ 1 & & & & \\ 1 & & M & & \\ \vdots & & & & \\ 1 & & & & \\ 0 & & & & \\ 0 & & M' & & \\ \vdots & & & & \\ 0 & & & & \\ 0 & 0 & 0 & \cdots & 0 \end{pmatrix}$$

(i) Show that any two rows of N differ in at least $2(k - \lambda)$ places if and only if $k \leqslant 2\lambda + 1$ and $v \geqslant 3k - 2\lambda$. [H]

(ii) Use the usual relationships between the parameters to show that the two inequalities in (i) hold if and only if $k = 2\lambda + 1$ and $v = 4\lambda + 3$. [H]

(iii) If we use the rows of N as the codewords of a code, show that it will have the same error-correcting capabilities as the code using M if and only if the original design was of the form $(4n - 1, 4n - 1, 2n - 1, 2n - 1, n - 1)$.

16

Ramsey theory

Before we talk in general terms about 'Ramsey theory' we give three examples (without proof): the first is trivial but the others (even for these low numbers) are not.

Examples (i) *If each of the one-element subsets of $\{1, 2, 3, 4, 5, 6, 7\}$ is coloured red or green, then (no matter how that colouring is done) there will be a four-element subset of $\{1, 2, 3, 4, 5, 6, 7\}$ all of whose one-element subsets are the same colour.*

(ii) *If each of the two-element subsets of $\{1, 2, \ldots, 18\}$ is coloured red or green, then (no matter how that colouring is done) there will be a four-element subset of $\{1, 2, \ldots, 18\}$ all of whose two-element subsets are the same colour.*

(iii) *If each of the three-element subsets of $\{1, 2, \ldots, 21\}$ is coloured red or green, then (no matter how that colouring is done) there will be a four-element subset of $\{1, 2, \ldots, 21\}$ all of whose three-element subsets are the same colour.* □

In general, let c be a positive integer (like the four in the above examples) and let k be a positive integer less than c (like the one, two or three in those progressive examples). Then it turns out that there exists a positive integer R (like the 7, 18 or 21 above) such that if all the k-element subsets of $\{1, \ldots, R\}$ are coloured red or green then there is bound to exist a c-element subset of $\{1, \ldots, R\}$ all of whose k-element subsets are the same colour. This is a restricted two-colour version of 'Ramsey's theorem' first proved (for any number of colours) by F.P. Ramsey in 1930: in each case the lowest possible number R is a 'Ramsey number'. But even in this restricted form the result is rather indigestible, and very little is known about specific values of these Ramsey numbers.

Let us start by looking more carefully at the cases $k = 1$ and $k = 2$, as in the first two examples above. The case $k = 1$ is a simple extension of the pigeon-hole principle. For example, in (i) above we are merely saying that if we put the numbers 1–7 into two pigeon-holes ('red' and 'green') then one of those pigeon-holes must contain at least four numbers. The case $k = 2$ involves colouring two-element subsets and it can be thought of as colouring the edges of a graph. For example, colouring all the two-element subsets of $\{1, \ldots, 18\}$ is like colouring all the edges of the complete graph with vertex-set $\{1, \ldots, 18\}$. Example (ii) above is then saying that, if the edges of K_{18} are coloured red or green in any way, then within it there is bound to exist a K_4 with all its edges of one colour. In this way Ramsey's theorem in the case $k = 2$ can be thought of as a combinatorial aspect of graph theory.

In the first part of this chapter we shall consider this graph-theoretic version of Ramsey's theorem and deduce several other delightful related results: later in the chapter we shall look at some wider aspects of 'Ramsey theory', giving us some stunning results with which to finish the book.

Example *Show that it is possible to colour the edges of K_5 red or green without creating a K_3 of one colour. Show, however, that if the edges of K_6 are coloured red or green, then there is bound to be a K_3 within it with all its edges the same colour.*

Solution It is easy to colour the edges of K_5 'red' or 'green' whilst avoiding a K_3 of one colour:

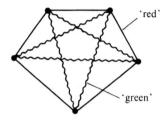

Now imagine that each of the edges of K_6 has been coloured red or green and consider any vertex of the graph. That vertex is the endpoint of five edges, each of which is red or green, and so at least three of those five edges are the same colour. Assume without loss in generality that at least three of them are red and consider the three vertices at the other ends of three red edges:

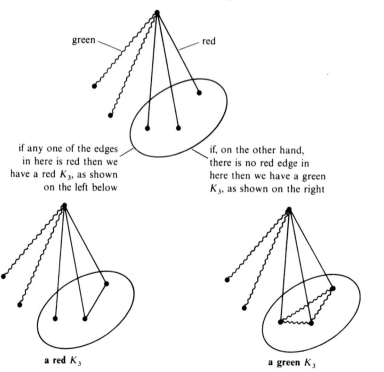

Ramsey theory

That example shows that $n = 6$ is the lowest number with the property that, if the edges of K_n are coloured red or green, then there is bound to exist a red K_3 or a green K_3. We now generalise that to higher numbers. For example what size complete graph has the property that, if you colour its edges red or green, then there is bound to be a K_{10} of one colour? It is by no means obvious that such a graph must exist because it might be possible to colour red or green the edges of arbitrarily large complete graphs whilst avoiding a K_{10} in one of the colours. However the next result shows, for example, that if K_{48620} has its edges coloured red or green then there is bound to be a K_{10} of one colour!

Theorem Let $r, g \geq 2$ be integers and let $n = \binom{r+g-2}{r-1}$. Then, if the edges of K_n are coloured red or green, there is bound to be a red K_r or a green K_g.

Proof The proof is by induction on $r + g$, the lowest possible case being when $r = g = 2$. In fact if either $r = 2$ or $g = 2$ then the result is trivial. For example, if $r = 2$ then $n = \binom{2+g-2}{2-1} = g$ and it is clear that if the edges of K_g are coloured red or green then there is bound to be either a red edge (which makes a red K_2) or, failing that, all the edges will be green and we will have a green K_g.

So assume that $r > 2$, $g > 2$ and that the result is known for lower $r + g$. Then in particular

(i) if $n_1 = \binom{r+(g-1)-2}{r-1}$ and the edges of K_{n_1} are coloured red or green then there is bound to be a red K_r or a green K_{g-1};

(ii) if $n_2 = \binom{(r-1)+g-2}{(r-1)-1}$ and the edges of K_{n_2} are coloured red or green then there is bound to be a red K_{r-1} or a green K_g.

Now to deduce the result for r and g, let $n = \binom{r+g-2}{r-1}$ and assume that the edges of K_n have been coloured red or green. Consider one particular vertex v and look at the edges which end there: there are $n - 1$ of them and, by the usual recurrence relation for the binomial coefficients, we have

$$n - 1 = \binom{r+g-2}{r-1} - 1 = \binom{r+g-3}{r-1} + \binom{r+g-3}{r-2} - 1$$

$$> \underbrace{\binom{r+(g-1)-2}{r-1} - 1}_{n_1 - 1} + \underbrace{\binom{(r-1)+g-2}{(r-1)-1} - 1}_{n_2 - 1}.$$

Therefore those $n - 1$ edges ending at v cannot consist of at most $n_1 - 1$ green edges *and* at most $n_2 - 1$ red edges: hence

either those $n - 1$ edges include at least n_1 green edges	*or* those $n - 1$ edges include at least n_2 red edges
In this case consider the complete graph K_{n_1} formed by n_1 vertices at the other ends of those n_1 green edges:	In this case consider the complete graph K_{n_2} formed by n_2 vertices at the other ends of those n_2 red edges:
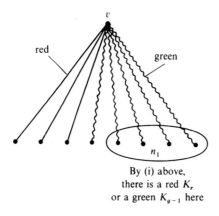	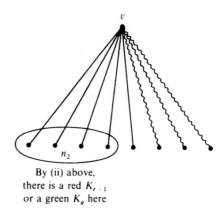
By (i) above, there is a red K_r or a green K_{g-1} here	By (ii) above, there is a red K_{r-1} or a green K_g here

There is a perfect symmetry between these two cases and so we consider, for example, the left-hand one. If there is a red K_r then we have found what we are looking for and the theorem is established. If on the other hand there is a green K_{g-1} in the subgraph shown, then overall with the extra vertex v there is a green K_g as required.

The right-hand case works similarly and therefore in each case there must be a red K_r or a green K_g within the coloured K_n, and the proof by induction is complete. □

Let $r, g \geq 2$. The previous theorem shows that there exists an integer n such that if the edges of K_n are coloured red or green then there is bound to be a red K_r or a green K_g: so we are now able to define the *Ramsey number* $R(r, g)$ as the smallest integer n with that property. Since the actual names of the colours are irrelevant it is easy to see that $R(r, g) = R(g, r)$. Our next theorem establishes some simple bounds on the Ramsey numbers.

Theorem *For $r, g \geq 2$ the Ramsey number $R(r, g)$ satisfies the inequalities*

$$(r-1)(g-1) + 1 \leq R(r, g) \leq \binom{r+g-2}{r-1}.$$

Ramsey theory

Proof The right-hand inequality is an immediate consequence of the previous theorem. For, by definition, $R(r, g)$ is the *smallest* integer n with the stated colouring property and, by that theorem, $n = \binom{r+g-2}{r-1}$ certainly has that property.

To establish the left-hand inequality we must simply exhibit a colouring of each of the edges of $K_{(r-1)(g-1)}$ in red or green for which there is no red K_r and no green K_g. To do this, draw the $(r-1)(g-1)$ vertices as an $(r-1) \times (g-1)$ rectangular grid. Then join any two vertices in the same row by a green edge and any two vertices in different rows by a red edge:

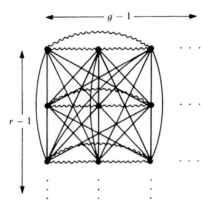

Then the green edges give a graph which consists of $r-1$ components, each of which is a complete graph on $g-1$ vertices. It is therefore easy to see that there is no K_g within this green graph (indeed there is no green connected subgraph on g or more vertices).

It is also easy to check that there is no red K_r within this colouring. For if we look at any r vertices they are shared out amongst the $r-1$ rows and so, by the pigeon-hole principle, one row must include at least two of those vertices. But then those two vertices are joined by a green edge and so they cannot form part of a red K_r.

Therefore we have coloured the edges of $K_{(r-1)(g-1)}$ red or green whilst avoiding a red K_r and a green K_g and it follows that $R(r, g) > (r-1)(g-1)$ as required. □

What are the actual numerical values of the Ramsey numbers? It is trivial to see that $R(2, g) = g$ for any $g \geq 2$: for, as we have already argued, if K_g has its edges coloured red or green then either there will be a red edge or, failing that, the whole of K_g will be green. (Alternatively, notice that the cases $r = 2$ and $g = 2$ are the only ones in which the two bounds in the previous theorem actually coincide to give an equality.) The example on page 202 shows that $R(3, 3) = 6$; as we shall see in the exercises, $R(3, 4) (= R(4, 3)) = 9$; it is also known that $R(3, 5) = 14$, $R(3, 6) = 18$ and $R(3, 7) = 23$. Also, as implied by part (ii) of the opening example of the chapter, $R(4, 4) = 18$. But apart from these no other actual values of the Ramsey numbers are known.

It is easy to extend this work from the case of two colours to m colours, and we just give one such result here:

Theorem *Let m be a positive integer. Then there exists an integer M such that if each edge of K_M is coloured in one of m colours there is bound to exist a K_3 in one colour.*

(In the exercises we shall see that $M = [m!\, e] + 1$ works.)

Proof The proof is by induction on m: the case $m = 1$ is trivial and the case $m = 2$ is part of the Ramsey theory established earlier (when we saw that if K_6 is coloured in two colours there will be a K_3 in one colour).

So assume that $m > 2$ and that the result is known for the case of $m - 1$ colours. Then by the induction hypothesis there exists a number M_0 such that, if the edges of K_{M_0} are coloured in $m - 1$ colours, there is bound to exist a K_3 of one colour. Let M be the Ramsey number $R(3, M_0)$. To show that M is as required in the theorem we shall assume that K_M has its edges coloured in m colours and we shall show that there is bound to be a K_3 in one colour.

Let one of the colours be red and for the moment describe all the remaining $m - 1$ colours as 'non-red'. Then, by the choice of M as the Ramsey number $R(3, M_0)$, there will be either a red K_3 or a non-red K_{M_0}. If there is a red K_3 then we have finished. If, on the other hand, there is a non-red K_{M_0} then its edges are coloured in $m - 1$ colours. Therefore, by the choice of M_0, this K_{M_0} (and hence the original K_M) will contain a K_3 of one colour. That completes the proof by induction. □

Since $R(3, 3) = 6$ and $R(3, R(3, 3)) = R(3, 6) = 18$, it is a consequence of that proof that if K_{18} has its edges coloured in one of three colours then there will be a K_3 of one colour. In fact we shall see in the exercises that the same is true of K_{17} (but it fails for K_{16}). The interested reader will be able to see how we could define a Ramsey number $R(r_1, \ldots, r_m)$, this example illustrating that $R(3, 3, 3) = 17$.

Another direction in which Ramsey theory can lead is to results where one is searching, in a large enough complete graph, not for a red K_r and a green K_g, but instead for some other prescribed red and green graphs. We now give one neat result of this type where, surprisingly, it is possible to be specific about the necessary size of the complete graph.

Theorem *Let $r, g \geqslant 2$ be integers and assume that we are given a particular tree T with g vertices. Then, if each edge of $K_{(r-1)(g-1)+1}$ is coloured red or green, there is bound to be a red K_r or a green T.*

Furthermore, $(r - 1)(g - 1) + 1$ is the lowest number with that property.

Proof Note firstly that the number $(r - 1)(g - 1) + 1$ certainly cannot be reduced. For, as commented at the time, the colouring of $K_{(r-1)(g-1)}$ given on the previous page has no red K_r and no connected green graph on g vertices (and hence no green T).

The proof that $(r - 1)(g - 1) + 1$ does work is by induction on $r + g$, the lowest possible case being when $r = g = 2$. In fact the result is trivial if either $r = 2$ or $g = 2$.

For example, if $r = 2$ then $n = (2 - 1)(g - 1) + 1 = g$ and, as before, it is clear that if the edges of K_g are coloured red or green then there will be either a red K_2 or a green K_g (which must include a green T).

So assume that $r > 2$, $g > 2$ and that the result is known for lower $r + g$. Before proceeding, picture the tree T which we are hoping to find in green. It has g vertices, and these must include a vertex v of degree 1 and an edge vw say. Let T' be the tree T with the vertex v and the edge vw removed: e.g.

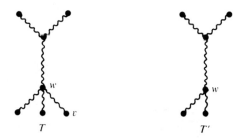

Now by the induction hypothesis the theorem holds for any lower value of $r + g$. In particular, by considering K_r and the tree T' on $g - 1$ vertices, we know that

(i) if $n_1 = (r - 1)((g - 1) - 1) + 1$ and the edges of K_{n_1} (or any bigger complete graph) are coloured red or green then there is bound to be a red K_r or a green T'.

Similarly, by the induction hypothesis applied to K_{r-1} and the tree T on g vertices, we know that

(ii) if $n_2 = ((r - 1) - 1)(g - 1) + 1$ and the edges of K_{n_2} are coloured red or green then there is bound to be a red K_{r-1} or a green T.

We now let $n = (r - 1)(g - 1) + 1$ and try to deduce the result for K_r and T within K_n. Note before proceeding that clearly $n > n_1$ and also that

$$n - (g - 1) = ((r - 1)(g - 1) + 1) - (g - 1) = (r - 2)(g - 1) + 1 = n_2.$$

So now assume that each edge of K_n is coloured red or green: we aim to show that there must be a red K_r or a green T.

Since $n > n_1$ it follows from (i) that the colouring will include either a red K_r (in which case we have finished) or a green T'. So assume the latter and locate one such copy of T' (on $g - 1$ vertices): then look at the remaining $n - (g - 1)$ $(= n_2)$ vertices.

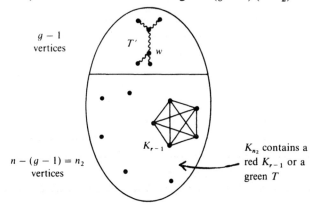

By (ii) this lower complete graph on n_2 vertices contains a red K_{r-1} or a green T. If it contains a green T then we have finished, so assume that it contains a red K_{r-1}, as shown.

Now consider the edges joining w to the $r-1$ vertices of that red K_{r-1}. If any one of them is green (as shown on the left below) we have found a green copy of the tree T. Failing that all those $r-1$ edges are red and we have found a red K_r (as shown on the right below).

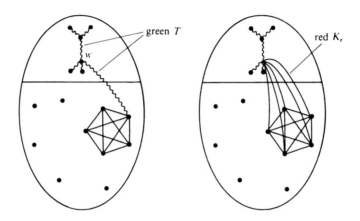

We have therefore deduced the result for r and g and the proof by induction is complete. □

By definition our graphs are always finite, and so in order to state an infinite version of Ramsey's theorem we return to its set-theoretic form. (But if you are willing to extend your concept of a graph to one with vertex-set $\{1, 2, 3, \ldots\}$ then you can continue to picture this result in a graph-theoretic way.)

Theorem *Let $N = \{1, 2, 3, \ldots\}$ and let each two-element subset of N be coloured in any one of a finite number of available colours. Then there exists an infinite subset S of N such that each two-element subset of S has the same colour.*

Proof Assume that each two-element subset of N has been coloured in any one of the finite number of available colours: we shall construct a set S as required.

In order to introduce a labelling which we can continue later, we let $N_0 = N$ and let n_0 be any particular member of N_0. Then consider the collection of two-element subsets $\{\{n_0, n\} : n \in N_0 \setminus \{n_0\}\}$. (For example, if we took n_0 to be 1 then this set of pairs would be simply $\{\{1, 2\}, \{1, 3\}, \{1, 4\}, \ldots\}$.) This infinite collection of pairs has been coloured using only a finite number of colours and so some infinite subcollection of them must be the same colour: assume that the infinite collection $\{\{n_0, n\} : n \in N_1\}$ are all the same colour (and, as an example, let us assume that all these pairs are red).

Now let n_1 be any member of N_1 and consider the collection of two-element subsets $\{\{n_1, n\} : n \in N_1 \setminus \{n_1\}\}$. This infinite collection of pairs has been coloured using only a finite number of colours and so some infinite subcollection of them must be the same colour: assume that the infinite collection $\{\{n_1, n\}: n \in N_2\}$ are all the same colour (and, as an example, let us assume that all these pairs are green).

Now let n_2 be any member of N_2 and consider the collection of two-element subsets $\{\{n_2, n\}: n \in N_2 \setminus \{n_2\}\}$. Again this collection has some infinite subcollection $\{\{n_2, n\}: n \in N_3\}$ of the same colour (blue, say).

Now let n_3 be any member of N_3 and consider ...

Continuing in this way gives us a sequence of infinite sets

$$N - N_0 \supseteq N_1 \supseteq N_2 \supseteq N_3 \supseteq \cdots$$

and a member of each of them

$$n_0 \in N_0, \ n_1 \in N_1, \ n_2 \in N_2, \ n_3 \in N_3, \ldots$$

such that for any particular n_i the two-element sets $\{\{n_i, n\}: n \in N_{i+1}\}$ are the same colour.

(If you allow infinite vertex-sets you can picture this process in a graph and compare it with our proofs of some of the earlier theorems:

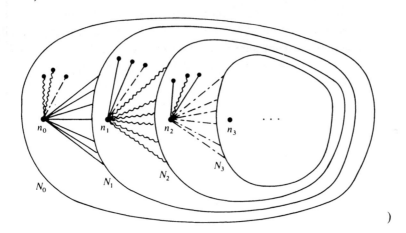

)

Therefore (with some specific colours used as an illustrative example) we have an infinite list

$$\{\{n_0, n\}: n \in N_1\} \text{ are all the same colour (red, say),}$$
$$\{\{n_1, n\}: n \in N_2\} \text{ are all the same colour (green, say),}$$
$$\{\{n_2, n\}: n \in N_3\} \text{ are all the same colour (blue, say),}$$
$$\{\{n_3, n\}: n \in N_4\} \text{ are all the same colour (red, say),}$$
$$\vdots$$

Since there is only a finite selection of colours available, some colour (yellow, say) must occur infinitely many times in that list; i.e.

$$\vdots$$
$$\{\{n_{i_1}, n\}: n \in N_{i_1+1}\} \text{ are all the same colour (yellow, say),}$$
$$\vdots$$
$$\{\{n_{i_2}, n\}: n \in N_{i_2+1}\} \text{ are all the same colour (yellow, say),}$$
$$\vdots$$
$$\{\{n_{i_3}, n\}: n \in N_{i_3+1}\} \text{ are all the same colour (yellow, say),}$$
$$\vdots$$

Let $S = \{n_{i_1}, n_{i_2}, n_{i_3}, \ldots\}$. Then S is infinite and we claim that each two-element subset of S is yellow. Consider a typical two-element subset $\{n_{i_j}, n_{i_k}\}$ with $n_{i_j} < n_{i_k}$. Then

$$n_{i_k} \in N_{i_k} \subseteq N_{i_k-1} \subseteq \cdots \subseteq N_{i_j+1}$$

and so the set $\{n_{i_j}, n_{i_k}\}$ is one of the pairs in $\{\{n_{i_j}, n\}: n \in N_{i_j+1}\}$ listed above and is therefore yellow, as claimed.

Hence each two-element subset of the infinite set S is yellow, and the theorem is proved. □

That theorem concludes our look at graph-theoretic aspects of Ramsey's theorem. As commented at the very beginning of the chapter, Ramsey's theorem actually extends to colouring three-element subsets and in general to colouring k-element subsets. We now prove a three-element form of the theorem and illustrate how the method can then be extended further.

Theorem *Let $r, g \geq 3$. Then there is a number n such that if each three-element subset of $V = \{1, 2, \ldots, n\}$ is coloured red or green, then either there exists an r-element subset of V all of whose three-element subsets are red or there exists a g-element subset of V all of whose three-element subsets are green.*

Proof The proof is by induction on $r + g$, the lowest case being when $r = g = 3$. But the result is trivial when $r = 3$ (or when $g = 3$) because we can then take n to be g (or r, respectively). So assume that $r > 3$, $g > 3$ and the result is known for lower $r + g$. Then in particular we can apply the induction hypothesis in the case of $r - 1$ and g to give a number n_1 with the property that

(i) if $|V_1| = n_1$ and each three-element subset of V_1 is coloured red or green, then either there exists an $(r - 1)$-element subset of V_1 all of whose three-element subsets are red or there exists a g-element subset of V_1 all of whose three-element subsets are green.

Similarly we can apply the induction hypothesis in the case of r and $g - 1$ to give a number n_2 with the property that

(ii) if $|V_2| = n_2$ and each three-element subset of V_2 is coloured red or green, then either there exists an r-element subset of V_2 all of whose three-element subsets are red or there exists a $(g-1)$-element subset of V_2 all of whose three-element subsets are green.

Now let R be the Ramsey number $R(n_1, n_2)$ and let $n = R + 1$. To show that n is as required in the theorem we assume that each three-element subset of $\{1, 2, \ldots, R+1\}$ has been coloured red or green and we find an r-element subset all of whose three-element subsets are red or a g-element subset all of whose three-element subsets are green.

Consider the complete graph K_R on vertex-set $\{1, 2, \ldots, R\}$ and colour each of its edges red or green using the same colour for the edge ij as was used for the three-element subset $\{i, j, R+1\}$. Then, by the definition of the Ramsey number $R = R(n_1, n_2)$, either there exists a K_{n_1} all of whose edges are red or there exists a K_{n_2} all of whose edges are green. Let us consider, for example, the existence of a K_{n_2} (as the other case is very similar).

The fact that there exists a K_{n_2} with all its edges green (and with vertex-set V_2, say) means that V_2 is a subset of $\{1, 2, \ldots, R\}$ with $|V_2| = n_2$ and such that each of the edges ij ($i, j \in V_2$) is green. By the way the colouring of the edges was defined, for V_2 this means that

> for each i, j in the set the three-element set $\{i, j, R+1\}$ is green.

Now by (ii) either there exists an r-element subset of V_2 all of whose three-element subsets are red or there exists a $(g-1)$-element subset W of V_2 all of whose three-element subsets are green. In the former case we have found an r-element subset of $\{1, 2, \ldots, R+1\}$ all of whose three-element subsets are red, and the theorem is established in this case. So assume that there exists a $(g-1)$-element subset W of V_2 such that

> each three-element subset of W is green.

So W has the properties (displayed above) that each three-element subset of W is green, and for each $i, j \in W$ the three-element set $\{i, j, R+1\}$ is green. It is therefore clear that each three-element subset of the g-element set $W \cup \{R+1\}$ is coloured green, and the proof by induction is complete. □

The interested reader will be able to see how to define a Ramsey number $R_k(r, g)$ involving the colouring of k-element subsets. Then our Ramsey number $R(r, g)$ is $R_2(r, g)$ and the proof of the previous theorem shows that

$$R_3(r, g) \leq 1 + R_2(R_3(r-1, g), R_3(r, g-1))$$

and enables the existence of $R_3(r, g)$ to be deduced by induction from the existence of the R_2s. Those with stamina will be able to verify that this extends to

$$R_4(r, g) \leq 1 + R_3(R_4(r - 1, g), R_4(r, g - 1))$$

and beyond, eventually showing the existence of all R_ks. In fact the full version of Ramsey's theorem actually encompasses k-element sets and m colours to give a number $R_k(r_1, \ldots, r_m)$.

We have now seen several results stemming from Ramsey's theorem. In very general terms that theorem says that, if you colour a large-enough collection, then some pattern must emerge within that colouring. For this reason 'Ramsey theory' now encompasses much more than his original theorem: it includes many non-trivial results which in some sense are saying that 'infinite disorder is impossible'. (We have in fact already met one such theorem in an earlier chapter: we saw on page 38 that in a sequence of $(r - 1)(g - 1) + 1$ distinct numbers there is bound to be an increasing subsequence of r terms or a decreasing subsequence of g terms.) The interested reader can find a very full account of 'Ramsey-type' theorems in the advanced text *Ramsey theory* by R.L. Graham, B.L. Rothschild and J.H. Spencer listed in the bibliography.

The rest of this chapter consists of a range of applications of Ramsey's theorem and its relatives.

Example *Show that if the numbers $\{1, 2, 3, 4, 5\}$ are each coloured red or green then there will exist x, y and z (not necessarily distinct) of the same colour and with $x + y = z$.*

Solution Of course, with such low numbers involved, it is easy to confirm this result by trial and error. However, in order to illustrate a general technique we shall derive this result from our earlier Ramsey theory. So assume that each number from 1 to 5 is coloured red or green: we shall construct a colouring of the edges of K_6 in red and green.

Given the complete graph K_6 on vertex-set $\{1, 2, 3, 4, 5, 6\}$ each of its edges can be given a number from 1 to 5 by the rule that the edge ij has number $|i - j|$. Then each edge can be coloured red or green according to the colour of its number from 1 to 5; e.g.

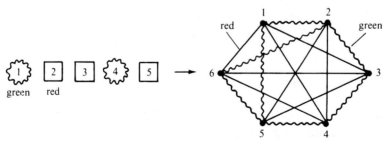

Note the red K_3 on vertices 1, 4 and 6
(since $2 = 6 - 4$, $3 = 4 - 1$ and $5 = 6 - 1$ are red)

Since the Ramsey number $R(3, 3)$ is 6 that K_6 is bound to include a K_3 of one colour: assume that it is on vertex-set $\{i, j, k\}$ with $i < j < k$ (like the one on vertex-set $\{1, 4, 6\}$ in the illustration). But this means that the edges with numbers $k - j, j - i$ and $k - i$ have the same colour. Therefore the numbers $x = k - j$, $y = j - i$ and $z = k - i$ have the same colour and satisfy $x + y = z$ as required. □

Our earlier theorem (page 206) concerning the colouring of K_M in m colours means that we can now trivially extend that example to the case of any number of colours: the proof follows the same lines as the previous solution and so it is left as an easy exercise.

Theorem (Schur) *Given a positive integer m let M be such that if the edges of K_M are coloured in m colours there is bound to exist a K_3 in one colour. (As we shall see in the exercises, $M = [m!\, e] + 1$ will do.) Then, if the numbers in $\{1, 2, \ldots, M - 1\}$ are each coloured in one of m colours, there are bound to exist x, y, z (not necessarily distinct) of the same colour and with $x + y = z$.* □

That result (which pre-dates Ramsey's theorem by over a decade) is named after the German mathematician I. Schur. However, his 1916 result actually concerns a form of 'Fermat's last theorem (modulo p)' and it uses the above Ramsey-type result in its proof. We state this theorem but omit some details of the proof because they require a knowledge of the theory of cosets.

Theorem *Given a positive integer m let M be as in the previous theorem and let p be any prime number with $p \geq M$. Then there exist $\alpha, \beta, \gamma \in \{1, 2, \ldots, p - 1\}$ with*

$$\alpha^m + \beta^m \equiv \gamma^m \pmod{p}$$

(and in fact α can be chosen to be 1).

Proof (with full details only for one example) Let m, M and p be as stated. Then, by the previous theorem, if each number in $\{1, 2, \ldots, p - 1\}$ is coloured in one of m colours, there will exist x, y, z (not necessarily distinct) of the same colour and with $x + y = z$.

Now the usual operations modulo p on $\{0, 1, \ldots, p - 1\}$ make it into a field and, in particular, each non-zero number x has a multiplicative inverse x^{-1}. Create a colouring of $\{1, 2, \ldots, p - 1\}$ such that x and y will have the same colour if and only if $x^{-1} \cdot y \equiv \delta^m \pmod{p}$ for some $\delta \in \{1, 2, \ldots, p - 1\}$. In general this works and it colours $\{1, 2, \ldots, p - 1\}$ in at most m colours. (Actually the non-zero mth powers form a multiplicative subgroup of $\{1, 2, \ldots, p - 1\}$ and the number of colours will be the number of cosets, namely the highest common factor of m and $p - 1$.) For example in the case $m = 3$ we could take $M = 18$ and $p = 19$. Then the only distinct non-zero mth powers modulo p are

$$1,\ 7\ (\equiv 4^3),\ 8\ (\equiv 2^3),\ 11\ (\equiv 5^3),\ 12\ (\equiv 10^3) \text{ and } 18\ (\equiv 8^3).$$

We now use these to create a colouring of $\{1, 2, \ldots, 18\}$ in three or fewer colours as described above. Note, for example, that in this field $6^{-1} = 16$ and therefore that $6^{-1} \cdot 4 \equiv 7$, which is one of those mth powers. Hence 6 and 4 will require the same colour. In fact it takes three colours to colour $\{1, 2, \ldots, 18\}$ in the stated way, one such colouring being:

    red: 1, 7, 8, 11, 12, 18   green: 2, 4, 6, 9, 10, 15   blue: 3, 5, 13, 14, 16, 17

(and algebraists will recognise these as the cosets with respect to the subgroup $\{1, 7, 8, 11, 12, 18\}$).

Now to return to the general case: this process colours $\{1, 2, \ldots, p-1\}$ in m colours and so by the earlier comments there will exist an x, y and z of the same colour and with $x + y = z$. Therefore in the field $\{0, 1, \ldots, p-1\}$ with operations modulo p we have

$$1 + x^{-1} \cdot y \equiv x^{-1} \cdot z \quad (\text{mod } p).$$

But

(i) $1 \equiv \alpha^m \pmod{p}$ where $\alpha = 1$,
(ii) as x and y are the same colour we have $x^{-1} \cdot y \equiv \beta^m \pmod{p}$ for some $\beta \in \{1, 2, \ldots, p-1\}$,
(iii) as x and z are the same colour we have $x^{-1} \cdot z \equiv \gamma^m \pmod{p}$ for some $\gamma \in \{1, 2, \ldots, p-1\}$.

Therefore

$$1 + x^{-1} \cdot y \equiv x^{-1} \cdot z \ (\text{mod } p) \quad \Rightarrow \quad \alpha^m + \beta^m \equiv \gamma^m \ (\text{mod } p)$$

for some $\alpha, \beta, \gamma \in \{1, 2, \ldots, p-1\}$ as required. □

Our next result from Ramsey theory has a similar framework to some earlier ones, but it is not actually a direct consequence of Ramsey's theorem.

Example *Show that if each of the numbers in $\{1, 2, \ldots, 325\}$ is coloured red or green then there will exist three different numbers x, y and z of the same colour and with $y = \frac{1}{2}(x + z)$.*

Solution Actually 325 is rather larger than necessary (in fact 9 will do!) but we are going to employ a method which generalises. Assume that we are given the numbers 1–325 each coloured red (R) or green (G). Our aim is to find three in the same colour such that one is exactly midway between the other two.

Write the numbers in a long line and divide them into 65 'boxes' of five:

    | 1 2 3 4 5 | | 6 7 8 9 10 | ⋯ | 161 162 163 164 165 | ⋯ | 321 322 323 324 325 |

Ramsey theory

Picture the colour pattern of each box: for example one of the boxes may have a pattern R R G G R and another may have pattern G R G G R. Altogether there are 32 ($= 2^5$) different possible colour patterns. Therefore (by the pigeon-hole principle) amongst the first 33 boxes two must have the same colour pattern: assume that the Ith and Jth boxes have the same pattern, where $1 \leqslant I < J \leqslant 33$. Then if $K = 2J - I$ it follows that $K \leqslant 65$ and so there is a Kth box in the list; furthermore the Jth box is exactly midway between the Ith and Kth boxes:

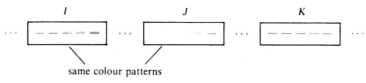

same colour patterns

three evenly spaced boxes

Now look at the first three entries in the Ith box: since there are only two colours these three entries must contain a repeated colour; assume that the ith and jth places are both red, say, where $1 \leqslant i < j \leqslant 3$. Then if $k = 2j - i$ it follows that $k \leqslant 5$ and so there is a kth place in that Ith box; furthermore the jth entry is midway between the ith and kth. We illustrate one such typical situation:

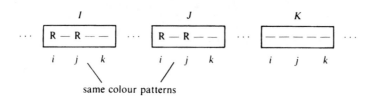

same colour patterns

What is the colour of the kth entry in the Ith box? If it is red then we immediately have three red numbers with one midway between the other two (namely the ith, jth and kth entries in the Ith box). So assume that the kth entry in the Ith box is green; since the Jth box has the same colour pattern, its kth entry is also green:

But now what is the colour of the kth entry in the Kth box (marked '?' above)? If it is green we can find three green numbers with one midway between the other two (namely the kth entries in the Ith, Jth and Kth boxes):

And if that kth entry in the Kth box is red we can find three red numbers with one midway between the others (namely the ith in the Ith box, the jth in the Jth box and the kth in the Kth box):

$$\cdots \boxed{\textcircled{R} - R - G} \cdots \boxed{R - \textcircled{R} - G} \cdots \boxed{- - - - \textcircled{R}} \cdots$$

So in all cases we have found three numbers x, y and z of the same colour with y midway between x and z; i.e. $y = \frac{1}{2}(x + z)$ as required. □

How was the number 325 chosen? Because there were two colours each 'box' had to contain $2 \cdot 2 + 1 = 5$ entries. That gave $2^5 = 32$ different colour patterns of boxes and so we needed $2 \cdot 32 + 1 = 65$ boxes; i.e.

$$325 = (2 \cdot 2 + 1) \cdot (2 \cdot 2^{2 \cdot 2 + 1} + 1) = (2 \cdot 2 + 1) \cdot (2 \cdot 2^{\text{previous bracket}} + 1).$$

Let us now consider the equivalent problem with three colours. Which L must we take to ensure that we can use a similar argument to show that if the numbers in $\{1, 2, \ldots, L\}$ are each coloured in red, green or blue then there are bound to exist three different numbers x, y and z of the same colour and with $y = \frac{1}{2}(x + z)$? As we shall see in the exercises an extension of the above argument does work with

$$L = (2 \cdot 3 + 1) \cdot (2 \cdot 3^{\text{previous bracket}} + 1) \cdot (2 \cdot 3^{\text{previous two brackets}} + 1)$$

and with four colours we need to take

$$L = (2 \cdot 4 + 1)(2 \cdot 4^{\text{previous one}} + 1) \cdot (2 \cdot 4^{\text{previous two}} + 1) \cdot (2 \cdot 4^{\text{previous three}} + 1) \, (!)$$

and so on. Hence, although the method does generalise to any number of colours, the actual sets become very large and the method rather technical: therefore we simply state the result for m colours without proof.

Theorem *Given a positive integer m there exists an integer L such that, if the numbers in $\{1, 2, \ldots, L\}$ are each coloured in one of m colours, then there are bound to exist three different numbers x, y and z of the same colour and with $y = \frac{1}{2}(x + z)$.* □

Another way of expressing the relationship between the three numbers in the theorem is to say that we are looking for different x, y and z of the same colour which form a three-term arithmetic progression. We now move up a stage to look for a four-term arithmetic progression in one colour.

Example *By the previous theorem there exists a number L such that, if the numbers in $\{1, 2, \ldots, L\}$ are coloured in 2^{650} colours, then there will be a three-term arithmetic progression in one colour. Show that, if the numbers in $\{1, 2, \ldots, 1300L\}$ are coloured in red and green, then there is bound to exist a four-term arithmetic progression in one colour.*

Solution Assume that we are given the numbers $1-1300L$ each coloured red (R) or green (G). Write the numbers in a long (!) list and divide them into $2L$ boxes each containing 650 numbers. The number of possible colour patterns of the boxes is 2^{650} and so, regarding these as ways of colouring the boxes, in the first half of the boxes we have a row of L each of which can be coloured in one of 2^{650} 'colours'. So by the choice of L we can apply the previous theorem to the *boxes* to show that there exist three of these first L boxes of the same colour pattern and with one box midway between the other two: assume that these are the Hth, Ith and Jth boxes with $H < I < J$, and picture these three boxes together with the Kth, where $K = 2J - I$:

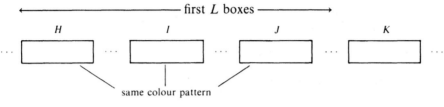

four evenly spaced boxes

As the Hth box contains 650 numbers it follows from the previous example that the 325 numbers in its first half will include a three-term arithmetic progression of one colour. Assume that the Hth box (and hence the Ith and Jth too) has such a red three-term arithmetic progression as shown:

If that three-term arithmetic progression continues to a fourth red term then we have found what we are looking for, so assume that the next term (which will still be in the same box) is green and then look at the corresponding position in the Kth box:

If that position (marked '?') is green then we can pick out the following four-term arithmetic progression in green:

If on the other hand that position marked '?' is red then we can pick out the following four-term arithmetic progression in red:

Hence in each case we have found a four-term arithmetic progression in one colour, and the solution is complete. □

The numbers in that example were chosen for convenience but are in fact much bigger than actually necessary. (The minimal ones which work are the *van der Waerden numbers* akin to our earlier Ramsey numbers.) That example shows how the theorem concerning three-term arithmetic progressions with any number of colours enables us (by applying it to some appropriately sized boxes) to deduce the equivalent result for a four-term arithmetic progression. The theorem for four-term progressions would then enable us to move up to five-term progressions, and so on. Without going into the rather horrendous details of this inductive proof, and without investigating any further the aspects of this topic, we state one general form of a theorem due to B. L. van der Waerden in 1927:

Theorem (van der Waerden) *Let r and m be positive integers. Then there is an integer L such that, if the numbers in $\{1, 2, \ldots, L\}$ are each coloured in one of m colours, then there is bound to exist an r-term arithmetic progression in one of the colours.* □

Our final application of Ramsey's theorem enables us to deduce a result about convex polygons. We start by recalling the terminology of the subject, most of which you will probably find familiar. A set in a plane is *convex* if for any two points in the set the line segment joining them is also in the set. Given any set in a plane its *convex hull* is the intersection of all convex sets containing it (and is hence the smallest convex set containing it). A *convex polygon* P is the convex hull of a finite set of points in a plane, and a *vertex* of P is a point $v \in P$ such that $P \setminus \{v\}$ is convex. E.g.

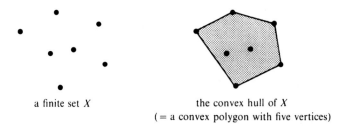

a finite set X the convex hull of X
(= a convex polygon with five vertices)

In order to deduce a Ramsey-type result about convex polygons we need two preliminary properties of them.

Example *Show that if five points are chosen in a plane, with no three in a line, then some four of them will form the vertices of a convex quadrilateral.*

Solution Consider the vertices of the convex hull of the five points: there will be three, four or five of them. E.g.

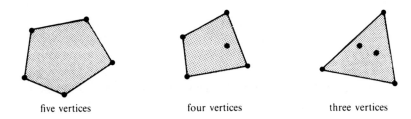

five vertices   four vertices   three vertices

In the case of five vertices *any* four will form the vertices of a convex quadrilateral, and in the case of four vertices those four are immediately the vertices of a convex quadrilateral. Finally, in the case when there are just three vertices, A, B and C, the two remaining points D and E will be inside the triangle formed by those three. The line through D and E meets just two sides of the triangle ABC: assume that it does not meet the side BC (as shown). Then it is easy to see that B, C, D and E form the vertices of a convex quadrilateral. □

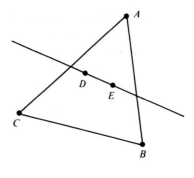

Example *Show that, if $r (\geq 4)$ points in a plane do not form the vertices of a convex polygon, then some four of the points do not form the vertices of a convex quadrilateral.*

Solution Assume that the r points do not form the vertices of a convex polygon. Then one of them, A say, is in the convex hull of the remaining $r - 1$. That convex hull will have as its vertices some of the r points (and we illustrate just these). Then A will be inside one of the triangles formed by three of those vertices, B, C and D say (as shown) or on BC say. But then the four points A, B, C and D do not form the vertices of a convex quadrilateral, as required. □

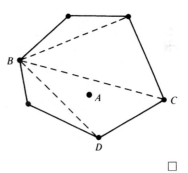

That very brief introduction to convex polygons is to enable us to deduce the following Ramsey-type theorem.

Theorem *Let r be a positive integer. Then there exists an integer R such that any R points in a plane with no three in a straight line must include r points which form the vertices of a convex polygon.*

Proof On page 212 we showed how to establish the existence of Ramsey numbers in the case of red and green four-element subsets. Therefore we know that if we are given any integers $r, g \geq 4$ then there is an integer $R = R_4(r, g)$ with the following property: if all the four-element subsets of an R-element set are coloured red or green there will exist an r-element subset all of whose four-element subsets are red or there will exist a g-element subset all of whose four-element subsets are green.

Now to prove this current theorem note firstly that it is trivial if $r \leq 3$. For then r points (with no three in a straight line) already form the vertices of a convex polygon and so we can take $R = r$.

So assume that $r \geq 4$ and let R be the Ramsey number $R_4(r, 5)$: to show that R is as required in the theorem we assume that we are given R points in a plane (with no three in a straight line) and we prove that there exist r of them which form the vertices of a convex polygon.

Colour each four-element subset of the R points as follows:

'red'	'green'
if those four points form the vertices of a convex quadrilateral;	if those four points do not form the vertices of a convex quadrilateral (i.e. if one of the points is in the convex hull of the remaining three).

Then, by the choice of R as the Ramsey number $R_4(r, 5)$, either there will exist

(i) r of the points such that any four-element subset of them is red; or
(ii) five of the points such that any four-element subset of them is green.

The first of our two previous examples on convex polygons shows that any five of these points will include four which form the vertices of a convex quadrilateral. This can be rephrased in terms of our colours by saying that any five of these points have a red four-element subset, and hence case (ii) above is impossible. Therefore (i) holds and there exist r of the points such that each four-element subset of them is red.

Hence we have found r points, any four of which form the vertices of a convex quadrilateral. By the second of the two examples above this means that the r points themselves must form the vertices of a convex polygon, as required, and the theorem is proved. □

That result brings us to the end of our work on Ramsey theory, and indeed nearly to the end of the book. It is appropriate that we have ended with an exciting topic which combines two of our major themes, selection arguments and graph theory. It is also appropriate that (as with so many of our other topics) the applications should cut across number theory, algebra and geometry and that there should be many avenues for further investigation (including the extensive exercises below). I hope that you have enjoyed this wide-ranging text showing combinatorial mathematics at work and that, for those of you who are going to continue studying mathematics, it has whetted your appetite for more.

Exercises

1. Show that, if each edge of K_6 is coloured red or green, then there will be at least two single-coloured K_3s. Show that, if K_7 is coloured in that way, then there will be at least three single-coloured K_3s.

2. Let q be a positive integer, let $Q = \binom{2q}{q}$ and assume that each edge of K_Q has

Ramsey theory

been coloured red or green so that there is no K_{q+2} of one colour. Show that there is a K_{q+1} of one colour and a K_q of the other colour. [H]

3. Show that for integers $r, g > 2$ the Ramsey numbers satisfy the inequality
$$R(r, g) \leq R(r-1, g) + R(r, g-1).$$
[H]

Show also that if both $R(r-1, g)$ and $R(r, g-1)$ are even then this inequality is strict. [H]

4. Let $g \geq 2$ be an integer and consider the complete graph on vertex-set $\{1, 2, \ldots, 3g - 4\}$. Colour an edge ij of this graph red if $|i - j| \equiv 1 \pmod{3}$ and green otherwise. Show that this coloured graph has no red K_3 and no green K_g. Deduce that $R(3, g) \geq 3(g - 1)$. [H]

5. Use exercises 3 and 4 to show that $R(3, 4) = 9$.

6. Let $r \geq 2$ be an integer. Show that $R(2r, 2r) > 2^r$. [H]

7. Given integers $r, g, b \geq 2$ let $R(r, g, b)$ be the lowest integer n such that, if K_n has its edges coloured red, green or blue, then there is bound to exist a red K_r, a green K_g or a blue K_b. Show that

(i) $R(r, g, 2) = R(r, g)$;

(ii) $R(r, g, b) \leq \min\{R(r, R(g, b)), R(g, R(r, b)), R(b, R(r, g))\}$; [H]

(iii) for $r, b, g > 2$
$$R(r, g, b) < R(r-1, g, b) + R(r, g-1, b) + R(r, g, b-1);$$
[H]

(iv) $R(r, g, b) \leq \dfrac{(r + g + b - 3)!}{(r-1)!\,(g-1)!\,(b-1)!}$. [H]

Use parts (i) and (iii) to deduce that $R(3, 3, 3) \leq 17$.

8. Let the sequence of integers $M_1, M_2, M_3, \ldots$ be defined by the recurrence relation
$$M_1 = 3 \quad \text{and} \quad M_m = mM_{m-1} - m + 2 \quad (m > 1).$$

(i) Show that $M_m = [m!\,e] + 1$. [H]
(ii) Show by induction that if the edges of K_{M_m} are coloured in m colours then there will exist a K_3 in one colour. [H]

9. Let $r \geq 2$ be an integer and define a sequence $R_1, R_2, R_3, \ldots$ by the recurrence relation
$$R_1 = r \quad \text{and} \quad R_m = \binom{2R_{m-1} - 2}{R_{m-1} - 1} \quad (m > 1).$$

Show that if the edges of K_{R_m} are coloured in 2^m colours then there will exist a K_r in one colour. [H]

10. Prove that if each edge of $K_{(r-1)^2+1}$ is coloured in red or green then either every tree on r vertices can be found in red or every tree on r vertices can be found in green.

11. Let G_1 be a graph with vertex-chromatic number $\chi(G_1) = r$ and let G_2 be a connected graph with g vertices. Find a colouring of the edges of $K_{(r-1)(g-1)}$ in red and green so that there is no red G_1 and no green G_2. [A]

12. Let T_1 be a tree with r vertices and let T_2 be a tree with g vertices in which there is a vertex of degree $g - 1$. Show that, if each edge of K_{r+g-2} is coloured red or green, then there is bound to exist a red T_1 or a green T_2. [H]
    In the case when $g - 2$ is a multiple of $r - 1$ show that $r + g - 2$ is the lowest number with that property, by exhibiting a colouring of K_{r+g-3} in red and green which avoids both a red T_1 and green T_2. [A]

13. Let us call a positive integer an *oddsum* if the sum of its prime factors is odd. (So that, for example, 54 is an oddsum since $54 = 2 \cdot 3 \cdot 3 \cdot 3$ and $2 + 3 + 3 + 3$ is odd.) Show that there exists an infinite set of positive integers such that the sum of any two of them is an oddsum. [H]

14. Assume that each positive integer has been coloured red or green. Now each rational number x with $0 < x < 1$ can be uniquely expressed in its lowest form as m/n for relatively prime positive integers m and n. Hence each such rational number has one of four 'colour-types'; red/red, red/green, green/red or green/green. Show that there exists an infinite set of positive integers of one colour such that for each pair of them the smaller divided by the larger gives a rational number of the same colour-type.

15. Given a square matrix a *principal submatrix* is obtained from it by deleting any of its rows and the same corresponding columns. Let us call a square matrix *semi-constant* if every entry above its leading diagonal is one constant and if every entry below its leading diagonal is also a constant (but not necessarily the same one). Then, for example, the matrix shown on the left has a 3×3 principal submatrix which is semi-constant, as indicated:

$$\begin{pmatrix} 1 & 0 & 1 & 1 & 0 \\ 1 & 1 & 1 & 0 & 1 \\ 0 & 0 & 0 & 1 & 1 \\ 1 & 1 & 0 & 0 & 1 \\ 1 & 0 & 0 & 0 & 1 \end{pmatrix} \xrightarrow[\text{rows and columns}]{\text{delete 1st and 4th}} \begin{pmatrix} 1 & 1 & 1 \\ 0 & 0 & 1 \\ 0 & 0 & 1 \end{pmatrix}.$$

Let $r \geq 2$ be an integer. Show that there exists an integer R such that any $R \times R$ matrix of 0s and 1s has an $r \times r$ principal submatrix which is semi-constant. [H]

Ramsey theory

16. Show that if each number in $\{1, 2, 3, 4, 5, 6, 7, 8, 9\}$ is coloured red or green then there is bound to be a three-term arithmetic progression in one colour.

17. Let $L = 7 \cdot (2 \cdot 3^7 + 1) \cdot (2 \cdot 3^{7 \cdot (2 \cdot 3^7 + 1)} + 1)$. Show that, if each number in $\{1, 2, \ldots, L\}$ is coloured red, green or blue, then there is bound to exist a three-term arithmetic progression in one colour. [H]

18. Let r and m be positive integers. Use the given form of van der Waerden's theorem to prove that there is a number L^* such that, if each of the numbers in $\{1, 2, \ldots, L^*\}$ is coloured in one of m colours, then there exists an r-term arithmetic progression with each of its terms **and** its common difference all of the same colour. [H]

Hints for exercises

Chapter 1

1. The 'various methods' include
expanding in terms of factorials;
comparing the coefficients of x^k in $(1 + x)^n$ and in $x^n\left(1 + \dfrac{1}{x}\right)^n$;
regarding 'choosing k from n' as 'choosing $n - k$ to leave out'; etc.

2. This is equivalent to placing $k - 1$ barriers in the $n - 1$ possible gaps in the queue.

3. *Don't* be tempted to refer back to the equivalent example with non-negative x_i: instead notice that partitioning n in this way is equivalent to breaking the queue into batches in exercise 2.

4. (i) $x_1 + x_2 + \cdots + x_k = n$ (each x_i a non-negative integer)

   $\equiv (x_1 + 1) + (x_2 + 1) + \cdots + (x_k + 1) = n + k$ (each $(x_i + 1)$ a positive integer)

   $\equiv y_1 + y_2 + \cdots + y_k = n + k$ (each y_i a positive integer)

5. This is equivalent to having $n - k$ empty seats and choosing k positions from the available gaps (including the ends) in which to place a seat with someone sitting in it.

7. (i) Coefficient of x^n in $(1 + x)^{2n} = (1 + x)^n (1 + x)^n$.
   (ii) Choosing n people is equivalent to choosing r men and $n - r$ women for some r (which is the same as choosing r men to *include* and r women to *exclude*).
   (iii) There are $\dbinom{2n}{n}$ shortest routes from A to B in the grid illustrated. Each passes through just one of the points on the diagonal shown.

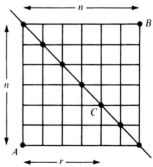

Hints for exercises

8. (iii) The various methods include a subset/leader argument again, or differentiating $(1+x)^n$.

9. (ii) The x_i lines meet each of the other $n - x_i$ lines, so the grand total of intersections is $\frac{1}{2}\sum x_i(n - x_i)$. You also know that $\sum x_i = n$.

   (iii) You need to find (by trial and error if all else fails) some x_i whose sum is 17 and whose squares add up to $17^2 - 2 \times 101$.

10. (i) Choosing a rectangle is equivalent to choosing two horizontal lines and two vertical lines.

    (ii) The number of rectangles required is the number in the top $r \times r$ corner but not in the top $(r-1) \times (r-1)$ corner.

11. (iii) How many of height n?   1
    How many of height $n-1$? $1 + 2$
    How many of height $n-2$? $1 + 2 + 3$
    $\vdots$

    (iv) How many of height 1? $1 + 2 + \cdots + (n-1)$
    How many of height 2? $1 + 2 + \cdots + (n-3)$
    $\vdots$

12. With no lines the circle has just one region. Now consider any collection of lines. If you draw a new line across the circle which does not cross any existing line then the effect is to increase the numbers of regions by 1:

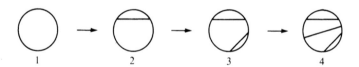

In addition, each time a new line crosses an existing line inside the circle the number of regions is increased by 1 again:

So in any such arrangement

$$\text{number of regions} = 1 + \text{number of lines} + \text{number of crossing points}.$$

13. (Case $m \geqslant n$.) Consider an $m \times n$ grid as shown. If we think of an 'up' as a 50p customer and a 'right' as a £1 customer then each of the usual routes from A to B corresponds to a possible arrangement of the queue.

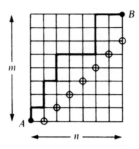

How many of those queues/routes would make the stall-keeper run out of change?

To count the routes from A to B which use at least one ringed junction, consider any such route and let R be the first ringed junction it uses. Then reflect that part of the route from A to R in the line of ringed junctions, as illustrated:

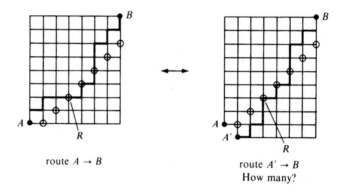

route $A \to B$ route $A' \to B$
How many?

(See also exercise 10 on page 125.)

Chapter 2

1. (ii) In how many ways can you choose m from the edges of K_n?
   (iii) How many subsets are there of the set of edges of K_n?

2. Count the ends of the edges. For a more formal argument use induction on the number of edges. (This result is called 'the handshaking lemma' because it says, in effect, that if some handshaking takes place then the grand total amongst everyone involved of all the hands shaken equals twice the number of handshakes that actually took place.)

4. (i) Induction on the number of vertices. (If $|V| > 1$ then there is a vertex of degree 1 whose removal gives a smaller tree.)
   (ii) Assume that the connected graph $G = (V, E)$ has $|E| = |V| - 1$ but that G is not a tree. Then G has a cycle and the removal of an edge of the cycle will leave a connected

Hints for exercises

graph. If this new graph still has a cycle, then repeat the process, etc. After the removal of some k (>0) edges you will obtain a tree. Why does that contradict (i)?

5. (ii) The degrees will be $n - 2, 2, 1, \ldots, 1$. So for $n > 4$ you have to choose a unique vertex i to have degree $n - 2$, a unique vertex j to which i is not joined, and a vertex k to which j is joined.

(iii) The tree must look like

(iv) Which vertex is 1 joined to? In how many ways can the tree on vertex set $\{2, \ldots, n\}$ be chosen?

For the last part you may use the fact that $\left(1 - \dfrac{1}{n}\right)^n \to \dfrac{1}{e}$.

6. (i) In C_nH_{2n+2} there are $3n + 2$ vertices, n of which have degree 4, the rest having degree 1. So what is the sum of the degrees and how many edges does this connected graph have? Then use exercise 4 (ii) to show that the molecular structure is tree-like.

For C_nH_{2n} this vertex and edge count show that it is not tree-like.

(ii) In C_4H_{10} the four carbon atoms and their joining edges form a tree on four unlabelled vertices. As we have seen, there are only two such trees, namely

and

Similarly the carbon atoms in C_3H_8 and C_5H_{12} form trees on three and five unlabelled vertices.

7. For the right-hand inequality show that, if two of the components are complete graphs on m and n vertices with $2 \leq m \leq n$, then there are more edges in complete components of $m - 1$ and $n + 1$ vertices. So how can you maximise the number of edges when the number of vertices and number of components are given?

Chapter 3

2. (i) Assume that $x, y \in B_1$, say, with $x \neq y$ and deduce a contradiction as follows. The families

$$(B_1 \backslash \{x\}, B_2, \ldots, B_n) \quad \text{and} \quad (B_1 \backslash \{y\}, B_2, \ldots, B_n)$$

fail to have the Hall-type property and so show that there exist $I_1, I_2 \subseteq \{2, \ldots, n\}$ with

$$\left|(B_1 \backslash \{x\}) \cup \left(\bigcup_{i \in I_1} B_i\right)\right| = |I_1| \quad \text{and} \quad \left|(B_1 \backslash \{y\}) \cup \left(\bigcup_{i \in I_2} B_i\right)\right| = |I_2|.$$

Deduce that

$$B_1 \setminus \{x\} \subseteq \bigcup_{i \in I_1} B_i \quad \text{and} \quad B_1 \setminus \{y\} \subseteq \bigcup_{i \in I_2} B_i \quad \text{and} \quad B_1 \subseteq \bigcup_{i \in I_1 \cup I_2} B_i.$$

You also know that

$$\left| \bigcup_{i \in I_1 \cap I_2} B_i \right| \geq |I_1 \cap I_2|.$$

Remembering that in general $|X \cup Y| = |X| + |Y| - |X \cap Y|$, show that the set $I = \{1\} \cup I_1 \cup I_2$ contradicts the fact that $\mathfrak{B}$ has the Hall-type property.

3. (i) *Start* our proof of Hall's marriage theorem by marrying B to one of the girls who knows him. In the subsequent method to construct fiancés for all the girls we reshuffled existing engaged couples, so at each stage B will be engaged to someone.
(ii) In the new situation there are m girls and m boys and it is straightforward to check that any r girls know between them at least r boys. So by Hall's theorem *all* the girls and boys can get married: who marries B?
(iii) The only way that the condition in the last theorem can fail in the case when $|P| = 1$ is when $|P \setminus (\bigcup_{i \in I} A_i)| = 1$ and $n - |I| = 0$, but that is impossible.

4. Invite an additional $n - k$ new very popular boys who are known by all the girls. Show that at least k girls can find husbands in the original situation if and only if *all* the girls can find husbands in the new situation. Then apply Hall's theorem to the new situation.

5. Induction on n. Given n girls satisfying the Hall-type conditions, consider the first girl. If she can marry any freely chosen boy from the ones she knows ($\geq m$ choices) and still leave a collection of unmarried girls and boys satisfying the Hall-type conditions, then induction easily gives the result. On the other hand, if by marrying one of the boys she knows she would leave a situation not satisfying the Hall-type conditions, then in the original situation there must be some n' ($m \leq n' \leq n$) girls who know between them precisely n' boys. In marrying off the n girls these n' must marry the n' boys they know and, by the induction hypothesis, their husbands can be chosen in at least $m!$ ways.

6. Invent m boys and n girls. Then think of a 1 in the ith row and jth column of M as meaning that the jth girl knows the ith boy.

7. Take any r sets in $\mathfrak{A}$. List all their elements (including repeats) giving at least rd entries in the list. No element occurs more than d times in that list and so there must be at least r *different* entries in the list.

10. To show that any doubly stochastic matrix M can be written as such a sum use induction on the number, γ, of non-zero entries in M, the lowest possible case being n. Now if $\gamma > n$ use the matrix form of Hall's theorem to show that in M there exists a set of non-zero entries with one in each row and one in each column, let λ_1 be the lowest of those particular

Hints for exercises

non-zero entries, let M_1 be the permutation matrix with 1s in the places of those particular non-zero entries, and apply the induction hypothesis to $\dfrac{1}{(1-\lambda_1)}(M-\lambda_1 M_1)$.

11. Define a bipartite graph $G=(V_1, E, V_2)$ with $V_1=\{x_1,\ldots,x_n\}$, $V_2=\{y_1,\ldots,y_n\}$ and with x_i joined to y_j if and only if $X_i \cap Y_j \neq \emptyset$. Then show that any r of the x_i are joined to at least r of the y_i and apply Hall's theorem.

(A normal subgroup of a finite group causes the group to be decomposed into its n left cosets, all of equal size, and also into its n right cosets, all of equal size. By exercise 11 there exist n elements in the group which simultaneously represent the left and right cosets.)

Chapter 4

2. (i) Use the shared pigeon-holes

□	□	□	...	□
1, 2n	2, 2n−1	3, 2n−2		n, n+1

(ii) Use shared pigeon-holes to show that two of the numbers are *consecutive*.

(iii) Use pigeon-holes

□	□	□	...	□
1	3	5		2n−1

with a number placed in the pigeon-hole corresponding to its largest odd factor.

3. With numbers $a_0, a_1, \ldots, a_n$ create pigeon-holes

□	□	□	...	□
0	1	2		n−1

and place each of the numbers $a_1 - a_0, \ldots, a_n - a_0$ in the pigeon-hole corresponding to the remainder when that number is divided by n.

4. (i) (ii)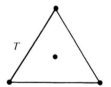

five points into four sets   four points into the three circles

5. Put the numbers $9, 99, 999, \ldots$ into the same pigeon-holes as in exercise 3 above. (Alternatively this can be proved by considering the decimal expansion of $1/n$: see, for example, chapter one of *Yet another introduction to analysis* listed in the bibliography.)

6. Dismiss the trivial case of $k = 0$. Then look at the total money after i days ($= t_i$ say). The $2n$ positive integers

$$t_1, t_2, \ldots, t_n, \quad \text{with} \quad k, k + t_1, k + t_2, \ldots, k + t_{n-1}$$

are all less than $2n$.

8. ($\Rightarrow$) Easy odd/even and black/white parity checks.

($\Leftarrow$) To show that if n is even and one black and one white square are removed then the remaining board can be covered, imagine the smallest rectangular part of the board containing the positions of the two removed squares. A parity check shows that this rectangle will be an even number of squares times an odd number:

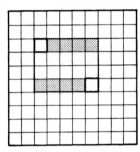

It is straightforward in general to fill in the shaded part, then the rest of the rectangle, and then the rest of the board.

10. Imagine each piece as a black 1 cm cube stuck to a white 1 cm cube.

11. (i) The Hamiltonian cycle will be of the form

$$\underbrace{\begin{array}{ccccc} v_1, & v_2, & v_3, & \ldots, & v_n, \\ \in V_1 & \in V_2 & \in V_1 & & \in V_2 \end{array}}_{\substack{\text{all the vertices are here} \\ \text{exactly once}}} \begin{array}{c} v_1 \\ \in V_1 \end{array}$$

(ii) The knight always moves from a black to a white square or vice versa.

15. (iii) Consider the n^m functions as the objects and let 'propery i' be that the function f never has any $f(j)$ equal to i. Then $N(1, 2, \ldots, r)$, for example, is the number of functions from $\{1, \ldots, m\}$ to $\{r + 1, r + 2, \ldots, n\}$ which, by (i), is $(n - r)^m$.

Chapter 5

2. Adding a next row to a $p \times n$ Latin rectangle with entries in $\{1, \ldots, n\}$ is equivalent to choosing distinct representatives, in order, from n sets each of size $n - p$: by exercise 5 of chapter 3 this can be done in at least $(n - p)!$ ways.

4. If $N \geq p + q$ then $L(i) \geq p + q - N$ for all $i \in \{1, \ldots, N\}$, but $N \geq n$ is clearly also needed.

Hints for exercises

6. (ii) Simply adapt the proof from pages 63–65 in the text. If, for example, $k(i - i')$ is divisible by n, where k and n are relatively prime, then $i - i'$ is divisible by n.

7. Assume there are p points. Then any particular point x can be paired with any of the $p - 1$ others and each such pair will lie in just one line. But also x is in $n + 1$ lines and is paired with n other points in each line. Now count, in two different ways, the grand total of pairs containing x throughout all the lines.

Since each point is in $n + 1$ lines and each line contains $n + 1$ points it is soon clear that the number of lines is also p.

Chapter 6

3. Induction on $p = \frac{1}{2}k$. If $p > 1$ form a new graph by removing the edges of a 'longest' path P (i.e. one with most edges) from one vertex of odd degree to another. Then apply the induction hypothesis on the non-trivial components of this new graph.

5. Use a parity argument to show that d is even. Show also that if vertices u and v are unjoined then there exists a vertex w with uw and vw in the edge-set. (In fact for $d > 0$ the graph G is also Hamiltonian but that is a much harder result due to C.StJ.A. Nash-Williams in 1969. This and related results are discussed in *Graph theory and related topics* edited by J.A. Bondy and U.S.R. Murty. For the easier fact that G is 'semi-Hamiltonian' see exercise 11.)

9. Let the n vertices of a graph be the n examinations and join two by an edge if they are set by a different examiner.

10. Consider any unjoined pair of vertices u and v:

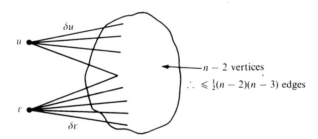

For the counter-example consider

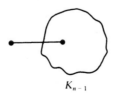

K_{n-1}

11. Show first that G is connected. Then let $|V| = n$ and let $v_1, \ldots, v_r$ be a longest path in G which does not use any vertex more than once. Assume that $r < n$ and deduce a contradiction by the following method. Note that v_1 is only joined to vertices from amongst $\{v_2, \ldots, v_r\}$ and that v_r is only joined to vertices from amongst $\{v_1, \ldots, v_{r-1}\}$. Also $\delta v_1 + \delta v_r \geq r$. So by an argument very similar to that in the proof on page 78 there is a cycle in G which uses all the vertices $v_1, \ldots, v_r$. That cycle, together with the connectedness of G, can be used to find a path in G consisting of $r + 1$ distinct vertices, thus contradicting the choice of r.

12. (ii) Assume that G is not complete and assume the weaker form of the given conditions, namely that $G\backslash\{u\}$ is always connected but that if the distinct vertices u, v and w have $uw, vw \in E$ and $uv \notin E$ then $G\backslash\{u, v\}$ is disconnected. Aim to show that for any three such vertices it follows that $\delta w = 2$ (and then that all the degrees are 2).

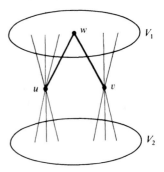

Let V_1 and V_2 be the sets of vertices into which $G\backslash\{u, v\}$ is disconnected and assume that $w \in V_1$, say, as illustrated. If $V_1 \neq \{w\}$ and x, y are as shown then $xu, yu \in E$ and $xy \notin E$ and so, by the given conditions, $G\backslash\{x, y\}$ is disconnected. The vertex u is in one of its components: let z be a vertex in its other component. As $G\backslash\{x\}$ and $G\backslash\{y\}$ are connected but $G\backslash\{x, y\}$ is not, there is a path in G from z to u which uses y but not x, there is one which uses x but not y, but all paths from z to u use x or y. Where is z?

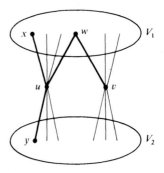

If it is in V_1 then the path from z to u avoiding x must use v, but then the path can go from z to v, w and u avoiding both x and y. Hence $z \notin V_1$. A similar argument shows that $z \notin V_2$. Deduce that $V_1 = \{w\}$ and that $\delta w = 2$.

Chapter 7

2. If the graph only has one vertex the two conclusions are immediate. Otherwise $d > 0$ and there is one vertex, v say, of degree less than d. Each of the d colours must be used at each vertex of degree d.
   (i) Consider one of the colours missing from v.
   (ii) Consider one of the colours (if any) used at v.

3. G is as illustrated: apply exercise 2 (and Vizing's theorem) to some appropriate subgraph of it.

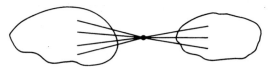

4. The graph must have an even number of vertices. Colour the edges of the Hamiltonian cycle alternately in two colours. How can you colour the remaining edges?

5. Edge-colour an appropriate bipartite graph: each colour gives some 'marriages'.

6. Associate the matrix with a bipartite graph $G = (V_1, E, V_2)$ with $|V_1| = m$ and $|V_2| = n$ and edge-colour the graph. Each coloured subgraph gives another matrix.

Chapter 8

2. Count the sets of three players in which player i won both games.

3. If the number of wins for player i is b_i, then the number of losses is given by $l_i = n - 1 - b_i$. But the losses $l_1, l_2, \ldots, l_n$ are the scores of another tournament.

4. (i) $\Rightarrow$ (ii). We know that the tournament has a directed path $p_1, \ldots, p_n$. In addition, if (i) holds there is an edge from p_n to some other p_i and so there is a directed cycle: let $p'_1, \ldots, p'_r, p'_1$ be the largest such. If there exists a player not used in this cycle, p say, then is it possible to slip her (and maybe others) somewhere in the cycle to give a bigger one $p'_1, \ldots, p'_{i-1}, p, p'_i, \ldots, p'_r, p'_1$?
(ii) $\Rightarrow$ (iii). If (ii) holds then in any proper subset of r players there will be at least one who has beaten one of the remaining $n - r$ players.
(iii) $\Rightarrow$ (i). If (i) fails the players can be partitioned into two sets P_1 and P_2 so that no player in P_1 beats any player in P_2. But then the scores of the players in P_1 will not satisfy (iii).

6. One way is to use edge-colourings with different days represented by different colours.

Chapter 9

1. Let $V_1 = \{r_1, r_2, \ldots, r_m\}$ and $V_2 = \{c_1, c_2, \ldots, c_n\}$ and define an $m \times n$ matrix M of 0s and 1s by the rule that the (i, j)th entry is 1 if and only if r_i is joined to c_j in G. What does a matching in G correspond to in the matrix? What does a set of vertices which include an endpoint of each edge correspond to?

2. By exercise 1 we can count instead the minimum number of vertices which between them include at least one endpoint of each edge of G. But given $W \subseteq V_1$, the set $V_1 \setminus W$ is part of such a set of vertices if and only if the set also includes $j(W)$:

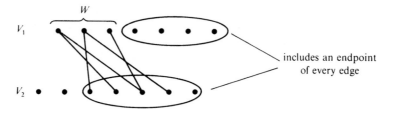

So if we are looking for a smallest such set we need only consider sets of the type $(V_1 \setminus W) \cup j(W)$ for $W \subseteq V_1$.

3. Show that, if an $m \times n$ matrix M of 0s and 1s has the Hall-type property that 'in any r rows there are 1s in at least r columns', then m is the smallest number of lines of M containing all its 1s. Hence use the König–Egerváry theorem to deduce the matrix form of Hall's theorem.

4. Given the graph $G = (V, E)$ define a network $N = (V, c)$ by the rule that $c(uv) \; (= c(vu))$ is 1 if u and v are joined in G and is 0 otherwise. Show that a minimum cut in N separating x from y will never use both a uv and a vu. Deduce that a minimum set of edges in G separating x from y corresponds to a minimum cut in N. Show also that a set of n edge-disjoint paths from x to y in G corresponds to a flow of value n in N which does not use both a uv and vu. Then apply the 'max flow–min cut' theorem to N.

Chapter 10

2. (i) How can the acceptable lists of length $n - 1$ be extended to acceptable lists of length n?

$c_n =$ (number of lists of length $n - 1$ ending in 0 with a 0, 1 or 2 tagged on)

+ (number of lists of length $n - 1$ ending in 1 with a 0 or 2 tagged on)

+ (number of lists of length $n - 1$ ending in 2 with a 0 or 1 tagged on).

And how many of the c_n lists end with a 0?

5. $y_1 + y_2 + \cdots + y_k = n$ if and only if $x^{y_1} x^{y_2} \cdots x^{y_k} = x^n$. So the required number of solutions is the number of x^ns in

$$(1 + x + x^2 + x^3 + \cdots)\underbrace{(1 + x + x^2 + x^3 + \cdots)}_{\text{(counting } y_2\text{)}} \cdots \underbrace{(1 + x + x^2 + x^3 + \cdots)}_{\text{(counting } y_k\text{)}}$$

$\underbrace{}_{\text{(counting } y_1\text{)}}$

6. (i) This is similar to the hint given for exercise 5 except that each time there is an n made

Hints for exercises

up of positive integers $y_1 + y_2 + \cdots + y_k$ there are precisely $y_1 y_2 \cdots y_k$ x^ns in the expansion:

$$\underbrace{(x + 2x^2 + 3x^3 + \cdots)}_{\text{(counting } y_1)} \underbrace{(x + 2x^2 + 3x^3 + \cdots)}_{\text{(counting } y_2)} \cdots \underbrace{(x + 2x^2 + 3x^3 + \cdots)}_{\text{(counting } y_k)}$$

(ii) Splitting the children into groups is like finding $y_1 + y_2 + \cdots + y_k = n$, and then choosing the leaders can be done in $y_1 y_2 \cdots y_k$ ways, so the problem reduces to that in (i). To illustrate one of the possible direct arguments consider, for example, seven children ($\times$) into three groups, with divisions between the groups marked | and with leaders circled; then replace the | and circled $\times$s by dashes:

e.g.
$$\times \ \circledtimes \ | \ \times \ \circledtimes \ \times \ | \ \circledtimes \ \times$$
$$\to \ \times \ - \ - \ \times \ - \ \times \ - \ - \ \times$$

With n children into k groups there is a one-to-one correspondence between the choice of groups/leaders and lists of $n - k$ $\times$s and $2k - 1$ dashes in any order whatever.

7. It is clear that for $i < j$ the (i,j)th entry of MN is zero. For $i \geqslant j$ that entry can be rewritten as

$$\binom{i}{i-j}\binom{j}{0} - \binom{i}{i-j-1}\binom{j+1}{1} + \binom{i}{i-j-2}\binom{j+2}{2} - \cdots + (-1)^{i-j}\binom{i}{0}\binom{i}{i-j}$$

which is the coefficient of x^{i-j} in

$$\underbrace{\left(\binom{i}{0} + \binom{i}{1}x + \binom{i}{2}x^2 + \cdots + \binom{i}{i-j}x^{i-j} + \cdots\right)}_{(1+x)^i}$$

$$\times \underbrace{\left(\binom{j}{0} - \binom{j+1}{1}x + \binom{j+2}{2}x^2 - \binom{j+3}{3}x^3 + \cdots\right)}_{(1+x)^{-(j+1)}}$$

And what *is* the coefficient of x^{i-j} in $(1 + x)^{i-j-1}$?

8. There are $\binom{n}{r} s_r$ functions from $\{1, 2, \ldots, m\}$ to $\{1, 2, \ldots, n\}$ whose image set contains precisely r elements. Then, writing out the given result as m equations, we get

$$\binom{1}{1}s_1 = 1^m,$$

$$\binom{2}{1}s_1 + \binom{2}{2}s_2 = 2^m,$$

$$\vdots \qquad \vdots \qquad \vdots$$

$$\binom{m}{1}s_1 + \binom{m}{2}s_2 + \cdots + \binom{m}{m}s_m = m^m.$$

Then use inversion.

10. How many of those upper routes touch the diagonal for the last time after $2r$ steps? There are u_r ways of reaching that particular last point on the diagonal and v_{n-r} ways of proceeding from there without touching the diagonal again. So there are $u_r v_{n-r}$ ($= u_r u_{n-r-1}$) such routes.

It follows that for $n > 1$ the coefficient of x^n in $u(x)$ equals the coefficient of x^{n-1} in $(u(x))^2$ and the required equation can be derived.

The quadratic equation in $u(x)$ leads to

$$u(x) = \frac{1 - (1 - 4x)^{1/2}}{2x}$$

and u_n, being the coefficient of x^n in this, is given by

$$-\tfrac{1}{2} \times \tfrac{1}{2} \times (-\tfrac{1}{2}) \times (-\tfrac{3}{2}) \times \cdots \times \left(-\frac{2n-1}{2}\right) \times (-4)^{n+1} \times \frac{1}{(n+1)!}$$

which tidies up.

12. The number of ways of scoring a total of n with two traditional dice is the coefficient of x^n in

$$g(x) = (x + x^2 + x^3 + x^4 + x^5 + x^6)^2 = x^2(1+x)^2(1-x+x^2)^2(1+x+x^2)^2.$$

If one of the new die has faces a, b, c, d, e and f (not necessarily distinct) and the other has faces a', b', c', d', e' and f' then the number of ways of scoring a total of n with the two new dice is the coefficient of x^n in

$$(x^a + x^b + x^c + x^d + x^e + x^f)(x^{a'} + x^{b'} + x^{c'} + x^{d'} + x^{e'} + x^{f'}).$$

Each of these factors has a zero constant term and has value 6 when $x = 1$. We must express $g(x)$ as the product of two such factors.

Chapter 11

3. Consider G vertex-coloured in $\chi(G)$ colours and consider any one of the colours. Show that there must exist a vertex in that colour which is joined to vertices of all the other colours.

6. For the first part let colours $1, 2, \ldots, \chi(G)$ vertex-colour G and let colours $1', 2', \ldots, \chi(\bar{G})'$ vertex-colour $\bar{G}$. Then use the $\chi(G) \cdot \chi(\bar{G})$ 'colours' $(1, 1'), (1, 2'), \ldots, (i, j'), \ldots, (\chi(G), \chi(\bar{G})')$ to vertex-colour the complete graph on vertex set V.

For the second part assume that the result fails and let $G = (V, E)$ be a smallest graph for which it fails. Let $v \in V$ and let G' be G with the vertex v (and v's edges) removed. Show that

$$\chi(G) = \chi(G') + 1, \qquad \chi(\bar{G}) = \chi(\overline{G'}) + 1 \quad \text{and} \quad \chi(G') + \chi(\overline{G'}) = |V|.$$

Show also that the degree of v in G is at least $\chi(G')$ and that its degree in $\overline{G'}$ is at least $\chi(\overline{G'})$. Deduce a contradiction.

Hints for exercises

9. Do *not* use the standard method based on the recurrence relation. Instead let G' be K_5 with edge uv removed. In how many ways can G' be vertex-coloured in k colours with u and v the same/different? Then piece two copies of G' together to give the graph G in question.

10. Let G_3 (G_4 respy) be the complete graph K_{n-1} (K_n respy) with one edge removed. Apply the standard recurrence relation (from the proof of the theorem on page 133) to the graphs G_1, G_2 and G_3 and hence express $p_{G_1} - p_{G_2}$ as the chromatic polynomial of a complete graph.

12. If $n > 1$ and the result is known for smaller graphs, then we know in how many ways the first $2(n-1)$ vertices can be properly coloured. Then for each such colouring in how many ways can the final two vertices be coloured
    (a) with u and v (as shown) of different colours, and
    (b) with u and v the same colour?

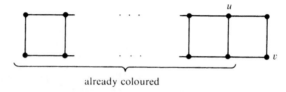

already coloured

13. (i) Apply the usual recurrence relation to G_1.
    (ii) Induction on n. If $n > 1$ then the process of (i) will give two graphs to which the induction hypothesis can be applied.

14. Induction on the number m of missing edges. If G has $m > 0$ missing edges and G_1, G_2 are as in the theorem on page 133, then the induction hypothesis can be applied to those graphs.

15. Check the result for the complete graphs (see exercise 5 on page 96) and then again use induction on the number of missing edges. Assume that G has missing edge uv and let G_1 and G_2 be obtained by the addition and contraction, respectively, of uv in the usual way. Show that each acyclic orientation of G is one of the following types:

   either the order $u \to v$ creates one acyclic orientation of G_1, and the ordering $v \to u$ creates another

   or just one of the orderings of uv creates an acyclic orientation of G_1.

   Show also that the number of acyclic orientations of G of the left-hand type equals the number of acyclic orientations of G_2 and deduce that

   $$\binom{\text{number of acyclic}}{\text{orientations of } G_1} - \binom{\text{number of acyclic}}{\text{orientations of } G_2} = \binom{\text{number of acyclic}}{\text{orientations of } G}.$$

Chapter 12

3. The required answer equals the number of ways of placing five non-challenging rooks on the unshaded board illustrated:

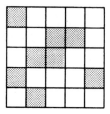

The *shaded* board's rook polynomial is easy to find (see answers).

4. Once again the required answer equals the number of ways of placing five non-challenging rooks on the unshaded board illustrated (and once again the *shaded* board's rook polynomial is easy to find):

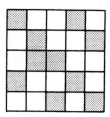

6. In this case

$$r_B(x) = 1 + r_1 x + r_2 x^2 + \cdots + r_{n-1} x^{n-1} + r_n x^n = r_{\bar{B}}(x)$$

and so

$$r_1 = \text{number of squares in } B = \text{number of squares in } \bar{B} = \tfrac{1}{2} n^2.$$

Also

$$\underbrace{n! - (n-1)! r_1 + (n-2)! r_2 - \cdots \pm 2! r_{n-2} \pm 1! r_{n-1}}_{\text{even}} \pm 0! r_n = r_n$$

7. Induction on n. If $n > 1$ and the result is known for lower n then

$$r_k = (\text{no. of ways of placing } k \text{ rooks where the last row is not used})$$
$$+ (\text{no. of ways of placing } k \text{ rooks where the last row is used})$$
$$= S(n, n-k) + (n-k+1) S(n, n-k+1).$$

8. Count by two different methods the number of ways of placing n non-challenging rooks on the $n \times n$ board with k or more in B and then colouring black some k of those in B.

Hints for exercises

9. (iii) Assume the first pack is order J1, J2, Q1, Q2, K1, K2, A1, A2 and label an 8 × 8 grid as shown:

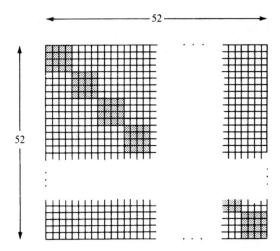

Then each of the 8! shufflings of the second pack corresponds to one of the ways of placing eight non-challenging rooks on this board; e.g. the arrangement shown corresponds to the cards being in the order K2, Q2, A1, J1, J2, K1, A2, Q1 which would give a 'snap' of Kings. The arrangements which avoid snaps are closely-related to those counted in (ii).

(iv) This time we have to count the number of ways of placing 52 non-challenging rooks on this unshaded part of the 52 × 52 board:

10. (i) It is possible to use induction on m and use the standard recurrence method to reduce the board of m squares to two boards of fewer squares. Alternatively, in the zigzag row of m squares, k rooks will be non-challenging if and only if no two of them are in adjacent squares: exercise 5 on page 10 counted such arrangements.

(ii) Use the standard recurrence method on this board with the nominated square 's' as the one in the bottom left-hand corner: then use (i).

(iii) Let wife 1 sit down first: she can choose any of the $2n$ seats. Then in how many ways can the remaining $n-1$ wives choose their seats? When they are all seated number the remaining seats as shown. In how many ways can the husbands now choose their seats? The board in (ii) is significant.

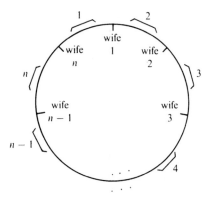

Chapter 13

2. In (ii) there are no cycles of three edges.

3. Consider *any* cycle C and assume that, in the given planar representation, it contains inside it precisely those faces bounded by the cycles $C_1, \ldots, C_r$ (each of which uses an even number of edges). Show that C uses an even number of edges.

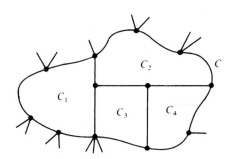

6. (ii) If $G = (V, E)$ and $\bar{G} = (V, \bar{E})$ are both planar with $|V| = n \ (\geq 3)$, then

$$|E| \leq 3n - 6 \quad \text{and} \quad |\bar{E}| \leq 3n - 6.$$

Adding these inequalities leads us to a quadratic inequality in n. Alternatively, rephrase this result in terms of thickness and use exercise 5. (In fact $t(K_n) = [\frac{1}{6}(n + 7)]$ for all positive integers n except $n = 9$ or 10. The proof that $t(K_9) > 2$ is quite technical and can be found, for example, in a paper by W.T. Tutte in the *Canadian Mathematical Bulletin*, volume 6, 1963.)

7. Note that there are n vertices on the perimeter of the figure (each of which has degree $n - 1$) and $\binom{n}{4}$ other vertices (each of which has degree 4): count the edges by summing the degrees.

8. (i) Let x be the number of faces bounded by four edges and y the number of faces bounded

Hints for exercises

by six edges. Then

$$x + y = |E| - |V| + 2;$$

$$2|E| = 4x + 6y \quad \text{(giving } |E| = 3(x + y) - x\text{);}$$

and

$$2|E| = 3|V|.$$

These equations have many solutions but in each of them $x = 6$.

10. Note that

$$|E| = \tfrac{1}{2} \sum_{v \in V} \delta v = \tfrac{1}{2} \sum_{n=0}^{\infty} n v_n \quad \text{and} \quad |V| = \sum_{n=0}^{\infty} v_n;$$

substitute these values into $|E| \leq 3|V| - 6$.

For the next part note that for connected graphs

$$5(v_1 + v_2 + v_3 + v_4 + v_5) \geq 5v_1 + 4v_2 + 3v_3 + 2v_4 + v_5 \geq \sum_{n=1}^{\infty} (6 - n)v_n \geq 12.$$

11. If $|V| > 6$ and the result is known for smaller graphs, then consider the graph obtained by the removal of a vertex of degree less than 6.

12. Let f be the number of faces. Then

$$f = |E| - |V| + 2 \quad \text{and} \quad 2|E| = cf = d|V|.$$

Divide the left-hand equation by $2|E|$.

13. The game would come to an end even if the planarity condition were dropped (although it would be a far less interesting game!). For if there are n vertices and m edges present, then what is the maximum number of edges which could be added to those vertices whilst giving no degree greater than 3? Consider the effect of the next turn and show that in every pair of turns the number of 'missing edges' reduces by 1.

Chapter 14

1. The inequalities

$$|E| - |V| + 2 \leq 11,$$

$$2|E| \geq 3|V|,$$

$$2|E| \geq 5(|E| - |V| + 2)$$

lead to a contradiction.

Use induction on the number of edges. In the inductive step consider a face bounded by three or four edges and reduce the map, as we did in the proof on page 171. A four-sided face can be reduced as in that proof and a three-sided face can be shrunk to a point.

2. In the proof by induction consider two coins which are furthest apart and show that each of those will be touching at most three other coins.

   To find an arrangement where three colours are not enough, note that in the layout

   if only three colours are used, then the two asterisked coins are forced to have the same colour. It is possible to string together some groups like this to lead to an arrangement where a fourth colour is needed.

3. Show that a function p exists if and only if the associated graph can be edge-coloured in three colours.

   ($\Rightarrow$) Assume p exists. Start at any edge and give it the label '1'. Then label all edges inductively (giving some edges several labels) by the rule that if e has a label l and e' is an edge which has a vertex v in common with e, then e' should be given a label as follows:

   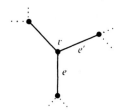   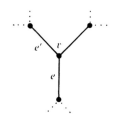

   $e \to e'$ clockwise around a face: $e \to e'$ anticlockwise around a face:
   give e' the label $l + p(v)$ give e' the label $l - p(v)$

   Show that this edge-colours the graph in the three 'colours'

   $$\{\ldots, -5, -2, 1, 4, 7, \ldots\}, \{\ldots, -4, -1, 2, 5, 8, \ldots\}, \{\ldots, -3, 0, 3, 6, \ldots\}.$$

   ($\Leftarrow$) Given an edge-colouring of the graph in three colours 1, 2 and 3, it follows that there is one edge of each colour at each vertex v. Define $p(v)$ by the following rule:

   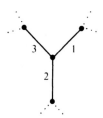   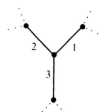

   $p(v) = 1$ if the colours 1, 2, 3 $p(v) = -1$ if the colours 1, 2, 3
   occur clockwise around v in the map   occur anti-clockwise around v in the map

4. Compare with exercise 6 of chapter 11 and imitate that solution. In particular, from the last theorem of the chapter, if $\bar{G}$ is planar then

   $$\chi(G \cup \bar{G}) \leq \chi(G) \cdot \chi(\bar{G}) \leq 5\chi(G).$$

Hints for exercises

Chapter 15

2. Take the new blocks to be the complements of the original ones and count the number of new blocks containing any pair of varieties.

3. (ii) Count, in two different ways, the total number of occurrences of members of B in the other blocks.
   (iii) Count, in two different ways, the total number of occurrences of pairs of members of B in the other blocks.

4. Use the usual relationships between the parameters to simplify $k + \lambda(v - 1)$. Then, if M is an incidence matrix of the design, we have

$$\underbrace{(\det M)^2 = \det(M^T M)}_{\text{square}} = \underbrace{(k + \lambda(v-1))}_{\text{square}} \underbrace{(k - \lambda)^{v-1}}_{\therefore \text{ square}}$$

5. Delete a whole line from a finite projective plane of order n.

6. If the process creates a design then (by the results on trees from chapter 2) three of the parameters will be

$$v = \tfrac{1}{2}n(n-1), \qquad b = n^{n-2}, \qquad k = n-1.$$

But then find λ and show that for $n > 3$ it cannot be an integer.

10. (i) Note that, for $n \geqslant 2$, given any $n \times n$ matrix M it follows that

$$\det\left(M + \begin{pmatrix} 1 & 0 & 0 & \cdots \\ 0 & 0 & 0 & \cdots \\ 0 & 0 & 0 & \cdots \\ \vdots & \vdots & \vdots & \end{pmatrix}\right) - \det\left(M - \begin{pmatrix} 1 & 0 & 0 & \cdots \\ 0 & 0 & 0 & \cdots \\ 0 & 0 & 0 & \cdots \\ \vdots & \vdots & \vdots & \end{pmatrix}\right) = 2 \det M',$$

where M' is the $(n-1) \times (n-1)$ matrix obtained by deleting the first row and column of M. It is now easy to establish the required result by induction on n.

(ii) Let Q be P with its last row and column deleted, let $\mathbf{p}$ be the last column of P with its last entry deleted, let D be as in (i) and solve the equations

$$(Q + D)\begin{pmatrix} x_1 \\ \vdots \\ x_{43} \end{pmatrix} = -\mathbf{p}.$$

(iv) If such rationals exist then there will be integers a, b and c with no common factor and with

$$6a^2 = b^2 + c^2.$$

But then, by considering the remainders when perfect squares are divided by 3, it is straightforward to deduce that b and c (and hence a) are divisible by 3.

(v) Note firstly that

$$P^T P = (\tfrac{1}{6}NL^T)(\tfrac{1}{6}LN^T) = \tfrac{1}{6}NN^T = \frac{1}{6}\begin{pmatrix} 7 & 1 & 1 & 1 & \cdots & 1 & 0 \\ 1 & 7 & 1 & 1 & \cdots & 1 & 0 \\ \vdots & \vdots & \vdots & \vdots & & \vdots & \vdots \\ 1 & 1 & 1 & 1 & \cdots & 7 & 0 \\ 0 & 0 & 0 & 0 & \cdots & 0 & 1 \end{pmatrix}$$

Finally use the fact that

$$6(y_1 \quad \cdots \quad y_{43} \quad y)\begin{pmatrix} y_1 \\ \vdots \\ y_{43} \\ y \end{pmatrix} = 6(x_1 \quad \cdots \quad x_{43} \quad 1)P^T P \begin{pmatrix} x_1 \\ \vdots \\ x_{43} \\ 1 \end{pmatrix}$$

and simplify these products, the right-hand one involving $(x_1 + \cdots + x_{43})^2$.

12. Let the b elements be in $k_1, \ldots, k_b$ of the sets and count the total occurrences of elements in pairs of sets to give

$$\tfrac{1}{2}v(v-1) = \sum_{i=1}^{b} \tfrac{1}{2}k_i(k_i - 1).$$

Then use the fact that the k_is sum to vr and that the sum of their squares is minimised when (and only when) each takes the average value vr/b. (In fact the notation has been chosen to illustrate the duality that in this case the incidence matrix of the sets is the transpose of an incidence matrix of a (v, b, r, k, λ) design.)

13. Let M be the $b \times v$ matrix whose rows consist of the b codewords. Show that any two rows of M both have 1s in $k - \tfrac{1}{2}d$ places and deduce that MM^T is the $b \times b$ matrix

$$\begin{pmatrix} k & k-\tfrac{1}{2}d & k-\tfrac{1}{2}d & \cdots & k-\tfrac{1}{2}d \\ k-\tfrac{1}{2}d & k & k-\tfrac{1}{2}d & \cdots & k-\tfrac{1}{2}d \\ k-\tfrac{1}{2}d & k-\tfrac{1}{2}d & k & \cdots & k-\tfrac{1}{2}d \\ \vdots & \vdots & \vdots & & \vdots \\ k-\tfrac{1}{2}d & k-\tfrac{1}{2}d & k-\tfrac{1}{2}d & \cdots & k \end{pmatrix}$$

with positive determinant. Now assume that $b > v$ and deduce a contradiction by adding $b - v$ columns of zeros to M and imitating the proof by contradiction of Fisher's inequality on pages 183–4.

14. (i) If the ith row ($i > 1$) has s 1s then the product of the ith row with the first shows that

Hints for exercises

$s - (m - s) = 0$. Also H and H^T commute in this case (H^T being a multiple of H^{-1}) and so a similar approach works for columns.

(ii) If the ith and jth columns ($1 < i < j$) have 1s together in t places, then use the product of the two columns to show that $t = m/4$.

15. (i) Compare all possible pairs of types of rows in N and see in how many places they differ. These are all at least $2(k - \lambda)$ if and only if

$$v - k \geqslant 2(k - \lambda), \qquad k + 1 \geqslant 2(k - \lambda), \quad \text{and} \quad 1 + v - 2(k - \lambda) \geqslant 2(k - \lambda),$$

which reduce to the two given inequalities.

(ii) Substituting $v = (k(k - 1)/\lambda) + 1$ into $v \geqslant 3k - 2\lambda$ and simplifying yields a quadratic inequality from which it follows that $k \geqslant 2\lambda + 1$.

Chapter 16

2. Note that $Q = \binom{(q+1)+(q+1)-2}{(q+1)-1}$ and so there will be a K_{q+1} in one colour, red say.
But then you can use the fact that $Q > \binom{(q+2)+q-2}{(q+2)-1}$.

3. Let $n = R(r - 1, g) + R(r, g - 1)$, assume that the edges of K_n have been coloured in red and green, and consider the $n - 1$ edges meeting at a particular vertex of K_n. There must be at least $R(r - 1, g)$ red edges or at least $R(r, g - 1)$ green and we then proceed in the usual fashion to find a red K_r or green K_g.

In the case when $R(r - 1, g)$ and $R(r, g - 1)$ are both even a quick parity check shows that in K_{n-1} (which has an odd number of vertices) it is impossible to have precisely $R(r - 1, g) - 1$ red edges meeting at each vertex.

4. It is easy to use modular arithmetic to show that there is no red K_3.

Assume that we have found a green K_p and (by symmetry) that 1 is one of its vertices. Then the other $p - 1$ vertices must be chosen from

$$3, 4, \quad 6, 7, \quad 9, 10, \quad \cdots \quad 3g - 9, 3g - 8, \quad 3g - 6, 3g - 5$$

with at most one from each of those pairs.

6. Consider a complete graph on a set of 2^r labelled vertices. There are $2^{\binom{2^r}{2}}$ ways of choosing a collection of red edges. Of those, $2^{\binom{2^r}{2} - \binom{2r}{2}}$ ways include the collection of $\binom{2r}{2}$ edges forming a complete graph on $2r$ particular vertices. So at most $\binom{2^r}{2r} 2^{\binom{2^r}{2} - \binom{2r}{2}}$ ways of choosing the red edges will include *some* red K_{2r}. At most twice that number of ways of

choosing the red edges will give a colouring containing a red K_{2r} or a green K_{2r}. Now show that, for $r \geqslant 2$,

$$2^{\binom{2^r}{2}} > 2\binom{2^r}{2r} 2^{\binom{2^r}{2} - \binom{2r}{2}}$$

and draw your own conclusions.

7. (ii) To show that $R(r, g, b) \leqslant R(r, R(g, b))$ imagine to start with that you are colour-blind between greens and blues.

(iii) If $n = R(r - 1, g, b) + R(r, g - 1, b) + R(r, g, b - 1) - 1$, then in K_n the edges meeting at any vertex must include either at least $R(r - 1, g, b)$ red edges or at least $R(r, g - 1, b)$ green edges or at least $R(r, g, b - 1)$ blue edges.

(iv) Use induction on $r + g + b$. The cases $r = 2$, $g = 2$ and $b = 2$ are straightforward and (iii) enables the induction to proceed.

8. (i) Show first that

$$[m!e] + 1 = 2 + m + m(m - 1) + m(m - 1)(m - 2) + \cdots + m!$$

It is then easy to show that this satisfies the given recurrence relation.

(ii) If K_{M_m} has its edges coloured in m colours then the $mM_{m-1} - m + 1$ edges meeting at a vertex must include M_{m-1} of one colour. Look at the ends of those edges and proceed in the usual way.

9. Assume that $m > 1$ and that the result is known for $m - 1$. If $R_m = \binom{2R_{m-1} - 2}{R_{m-1} - 1}$ and K_{R_m} has its edges coloured in 2^m colours, then regard those colours as 2^{m-1} shades of red and 2^{m-1} shades of green.

12. If $r > 2$, $g > 2$, and the result is known for lower $r + g$, then let T'_1 be T_1 with a vertex v of degree 1 (and its edge vw) removed. Then in any red and green coloured K_{r+g-2} there will exist a red T'_1 or a green T_2. In the former case look at all the edges in K_{r+g-2} which join w to the $g - 1$ vertices not already used in T'_1 (as shown).

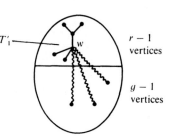

13. Colour the two-element subsets of $N = \{1, 2, 3, \ldots\}$ by the rule that $\{i, j\}$ is 'red' if $i + j$ is an oddsum (and 'green' otherwise). Then there exists an infinite set $S \subseteq N$ with each two-element subset of S the same colour: we hope that colour is red. If not, then try doctoring S in some way.

15. Take R to be $R(r, r, r, r)$, so that if the edges of K_R are coloured in four colours then there will exist a K_r of one colour. (The existence of such an R follows, for example, from

Hints for exercises

exercise 9.) Then given any $R \times R$ matrix (m_{ij}) of 0s and 1s define a colouring of each edge of K_R (on vertex-set $\{1, 2, \ldots, R\}$) in one of four colours in the following way: for $i < j$ the edge ij is coloured

$$\text{'red' if } m_{ij} = 0 \text{ and } m_{ji} = 0;$$

$$\text{'green' if } m_{ij} = 0 \text{ and } m_{ji} = 1; \text{ etc.}$$

17. (This is a similar argument to that used in the example on pages 214–6, but it continues for one extra stage.) Write 1–L in a long line, make little boxes each of seven numbers, and make larger boxes each of $2 \cdot 3^7 + 1$ little boxes:

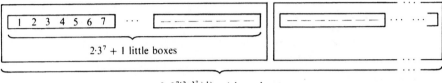

There are $3^{7(2 \cdot 3^7 + 1)}$ ways of colouring the larger boxes and so there will be a repeated pattern in the Ith and Jth large boxes for some I and J with $I < J \leqslant 3^{7(2 \cdot 3^7 + 1)} + 1$. If $K = 2J - 1$ then there will still be a Kth large box. Applying the fairly standard argument within the Ith large box gives, for example,

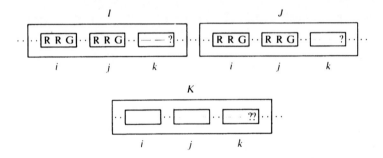

Consider the cases of '?' being red, green or blue (in the last case you will need to consider '??' too).

18. Induction on m. Assume that the result is known for $m - 1$ colours and let L be such that if $\{1, 2, \ldots, L\}$ is coloured in $m - 1$ colours then there will be an r-term arithmetic progression **and** common difference in one colour. Then (by van der Waerden's theorem) let L^* be such that if $\{1, 2, \ldots, L^*\}$ is coloured in m colours there is bound to exist an $((r - 1)L + 1)$-term arithmetic progression in one colour. Consider such a colouring and an $((r - 1)L + 1)$-term arithmetic progression in red say:

$$a, \quad a + d, \quad a + 2d, \quad a + 3d, \quad \cdots, \quad a + (r - 1)Ld.$$

Consider the two cases of whether jd is red for some j with $1 \leq j \leq L$ (in which case find the required red arithmetic progression with common difference jd) or whether all the numbers

$$d, \quad 2d, \quad 3d, \quad \cdots, \quad Ld$$

are in one of the $m-1$ non-red colours (in which case you can apply the induction hypothesis).

Answers to exercises

Chapter 1

2 and 3. $\binom{n-1}{k-1}$

5. $\binom{n-k+1}{k}$

9. (i) $\binom{n}{2}$

(iii) The only sets of positive integers which add to 17 and whose squares add to 87 are $\{1, 5, 5, 6\}, \{1, 2, 3, 3, 8\}, \{1, 1, 1, 2, 4, 8\}$ and $\{2, 3, 5, 7\}$. So one solution consists of two parallel lines in one direction, three in another, five in another and seven in another, as illustrated (17 lines, 101 intersection points).

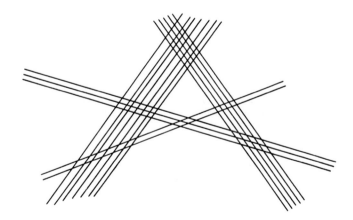

10. (i) $\binom{n+1}{2}^2$

(iv) $1^2 + 2^2 + \cdots + n^2$ (which equals $\frac{1}{6}n(n+1)(2n+1)$).

11. (i) $\binom{n+2}{3}$

(iii) $\binom{n+2}{3}$

(iv) $\binom{n}{2} + \binom{n-2}{2} + \binom{n-4}{2} + \cdots + \begin{cases} \binom{2}{2} & (n \text{ even}) \\ \binom{3}{2} & (n \text{ odd}) \end{cases}$

$(= \frac{1}{24}n(n+2)(2n-1)$ for n even and $\frac{1}{24}(n-1)(n+1)(2n+3)$ for n odd).

13. $1 - \dfrac{\binom{m+n}{m+1}}{\binom{m+n}{m}} = 1 - \dfrac{n}{m+1}$ $(m \geq n)$, 0 otherwise.

[In the case $m = n$ the number of acceptable queues is $\dfrac{1}{n+1}\binom{2n}{n}$, a Catalan number considered again in exercise 10 on page 125.]

Chapter 2

1. (i) $\binom{n}{2}$ or $\frac{1}{2}n(n-1)$

(ii) $\binom{\frac{1}{2}n(n-1)}{m}$

(iii) $2^{n(n-1)/2}$

5. (i) n (provided that $n > 2$) and just 1 in the case $n = 2$.
(ii) $n(n-1)(n-2)$ (provided that $n > 4$) and 12, 3, 0 in the cases $n = 4, 3, 2$ respectively.
(iii) $n!/2$
(iv) $(n-1)^{n-2}$

6. (ii) C_3H_8 is unique, namely

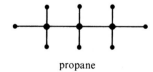

propane

Answers to exercises

There are three possible isomers of C_5H_{12} and in fact they all exist:

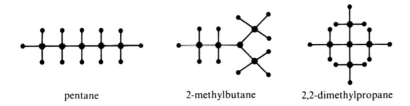

pentane   2-methylbutane   2,2-dimethylpropane

Chapter 3

1. (i) There are three ways of finding husbands for the girls: they can marry (in order) the boys 1, 4, 2, 3, 6, 5 or the boys 2, 4, 1, 3, 6, 5 or the boys 2, 4, 3, 1, 6, 5.
(ii) It is impossible to find husbands for all the girls. For example the girls 1, 2, 3 and 5 only know boys 1, 3 and 5 between them.

Chapter 4

11. (ii) One of the many possible knight's circular routes is

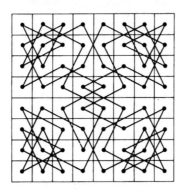

12. (i) 7334   ($= 5000 + 3333 + 2000 - 1666 - 1000 - 666 + 333$)

    (ii) 7715   ($= 5000 + 3333 + 2000 + 1428 - 1666 - 1000 - 666 - 714 - 476 - 285 + 333 + 238 + 142 + 95 - 47$)

13. Inclusion/exclusion gives

$$\binom{52}{5} - \binom{4}{1}\binom{48}{5} + \binom{4}{2}\binom{44}{5} - \binom{4}{3}\binom{40}{5} + \binom{4}{4}\binom{36}{5} = 10\,752.$$

14. $\dfrac{1}{2!} - \dfrac{1}{3!} + \dfrac{1}{4!} - \cdots + \dfrac{(-1)^n}{n!}$

$\left(= 1 - \dfrac{1}{1!} + \dfrac{1}{2!} - \dfrac{1}{3!} + \dfrac{1}{4!} - \cdots + \dfrac{(-1)^n}{n!} \text{ which tends to the series expansion of } e^{-1}. \right)$

15. (i) n^m
    (ii) $n(n-1)(n-2)\cdots(n-m+1)$

Chapter 5

1. i must be 6 and there is no possible value of j. One extension when $i = 6$ is

$$\begin{pmatrix} 1 & 2 & 3 & 4 & 5 & 6 \\ 5 & 6 & 1 & 2 & 4 & 3 \\ 3 & 4 & 5 & 1 & 6 & 2 \\ 4 & 1 & 2 & 6 & 3 & 5 \\ 2 & 3 & 6 & 5 & 1 & 4 \\ 6 & 5 & 4 & 3 & 2 & 1 \end{pmatrix}$$

4. $N = \max\{n, p+q\}$

5. $\begin{pmatrix} 1 & 2 & 3 & 4 \\ 4 & 3 & 2 & 1 \\ 2 & 1 & 4 & 3 \\ 3 & 4 & 1 & 2 \end{pmatrix}$

6. (i) The answer (with 1 added to each entry) is shown in the example on page 60.

(iii) $\begin{pmatrix} 0 & 1 & 2 & 3 \\ 1 & 0 & 3 & 2 \\ 2 & 3 & 0 & 1 \\ 3 & 2 & 1 & 0 \end{pmatrix}$ $\begin{pmatrix} 0 & 1 & 2 & 3 \\ 2 & 3 & 0 & 1 \\ 3 & 2 & 1 & 0 \\ 1 & 0 & 3 & 2 \end{pmatrix}$ $\begin{pmatrix} 0 & 1 & 2 & 3 \\ 3 & 2 & 1 & 0 \\ 1 & 0 & 3 & 2 \\ 2 & 3 & 0 & 1 \end{pmatrix}$

(Note that, as always with this type of construction, the first Latin square in (i) and (iii) consists of the addition table.)

Chapter 6

1. (i) K_n is Eulerian for n odd and it is semi-Eulerian for $n = 2$.

(ii) $K_{m,n}$ is Eulerian for m and n even and it is semi-Eulerian for $m = n = 1$ and for m odd with $n = 2$ (or vice versa).

(iii) K_n is Hamiltonian for $n \geq 3$.

(iv) $K_{m,n}$ is Hamiltonian for $m = n > 1$.

2. Routes exist only in the cases $n = 1, 2$ or 3 (for in the first two cases there are no possible moves!).

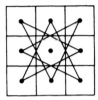

For $n \geq 4$ the graph of knight's moves is not Eulerian because it has eight vertices of odd degree; e.g.

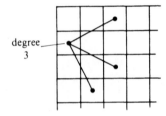

4. The smallest example consists of a cycle on five vertices.

Chapter 8

1. (i) The scores must be $n - 1, n - 2, \ldots, 2, 1, 0$ and there are $n!$ ways of deciding which player has which score. (The results are then determined as every game will have been won by the player with the higher score.)

(ii) None.

(iii) n must be even: the top score is then $n - 1$ and the other scores are all $\frac{1}{2}n - 1$.

2. $\binom{n}{3} - \binom{b_1}{2} - \binom{b_2}{2} - \cdots - \binom{b_n}{2}$.

6. (i) n days are needed if n is odd but $n - 1$ will do if n is even.

(ii) $2n$ days are needed if n is odd but $2(n - 1)$ will do if n is even.

Chapter 10

1. (i) $g_n = g_{n-1} + g_{n-2}$ with $g_1 = 2 (=F_3)$ and $g_2 = 3 (=F_4)$: so $g_n = F_{n+2}$.

   (ii) $j_n = j_{n-1} + j_{n-2}$ with $j_1 = 1 (=F_2)$ and $j_2 = 2 (=F_3)$: so $j_n = F_{n+1}$.

2. (ii) $c_n = \frac{1}{2}((1+\sqrt{2})^{n+1} + (1-\sqrt{2})^{n+1})$

   $= \binom{n+1}{0} + 2\binom{n+1}{2} + 2^2\binom{n+1}{4} + 2^3\binom{n+1}{6} + \cdots$

   (iii) $c_n =$ coefficient of x^n in $(1+x)(1+x(2+x)+x^2(2+x)^2+\cdots)$

   $= \left(\binom{n}{0}2^n + \binom{n-1}{1}2^{n-2} + \binom{n-2}{2}2^{n-4} + \cdots\right)$

   $+ \left(\binom{n-1}{0}2^{n-1} + \binom{n-2}{1}2^{n-3} + \binom{n-3}{2}2^{n-5} + \cdots\right)$

3. r_n is the coefficient of x^n in

   $$r(x) = \frac{x}{(1-x-2x^2)(1-x)} = \frac{x}{(1-2x)(1+x)(1-x)} = \frac{2}{3(1-2x)} - \frac{1}{6(1+x)} - \frac{1}{2(1-x)}$$

   which is

   $$(\tfrac{2}{3} \times 2^n) - \begin{cases}\tfrac{2}{3} \text{ if } n \text{ is even}\\ \tfrac{1}{3} \text{ if } n \text{ is odd}\end{cases} = \left[\frac{2^{n+1}}{3}\right],$$

   where $[\cdots]$ denotes the 'integer part of'.

6. (i) $\binom{n+k-1}{2k-1}$

8. Inversion of the equations in the s_n gives

   $\binom{1}{1}1^m = s_1,$

   $-\binom{2}{1}1^m + \binom{2}{2}2^m = s_2,$

   $\binom{3}{1}1^m - \binom{3}{2}2^m + \binom{3}{3}3^m = s_3,$

   $\vdots$

   and in general

   $s_n = n^m - \binom{n}{n-1}(n-1)^m + \binom{n}{n-2}(n-2)^m - \cdots + (-1)^{m-1}\binom{n}{1}1^m$

   as before.

Answers to exercises

11. (iii) A surjection from $\{1, 2, \ldots, m\}$ to $\{1, 2, \ldots, n\}$ is like partitioning $\{1, 2, \ldots, m\}$ into n sets where the order of the sets matters. Hence (from the answer to exercise 8)

$$S(m, n) = \frac{1}{n!}\left(n^m - \binom{n}{n-1}(n-1)^m + \binom{n}{n-2}(n-2)^m - \cdots + (-1)^{m-1}\binom{n}{1}1^m\right)$$

(iv) $M = \begin{pmatrix} 1 & 0 & 0 & 0 \\ -1 & 1 & 0 & 0 \\ 2 & -3 & 1 & 0 \\ -6 & 11 & -6 & 1 \end{pmatrix}$ $N = \begin{pmatrix} 1 & 0 & 0 & 0 \\ 1 & 1 & 0 & 0 \\ 1 & 3 & 1 & 0 \\ 1 & 7 & 6 & 1 \end{pmatrix}$

12. $(x + x^2 + x^3 + x^4 + x^5 + x^6)^2$

$$= (x(1+x)(1+x+x^2))(x(1+x)(1+x+x^2)(1-x+x^2)^2)$$
$$= (x + 2x^2 + 2x^3 + x^4)(x + x^3 + x^4 + x^5 + x^6 + x^8).$$

So the two new dice have faces 1, 2, 2, 3, 3, 4 and 1, 3, 4, 5, 6, 8. It is debatable which are better for *Monopoly*: I would prefer these new ones because 'doubles' are less frequent and I would go to jail less often!

Chapter 11

5. Let the vertices be $v_1, v_2, \ldots, v_n \; (= w)$ in any decreasing order of their distance from w. Let the colours be $1, 2, \ldots, d$. Colour the vertices in order, at each stage using the lowest-numbered colour for v_i not already used at a vertex adjacent to v_i. It is easy to check that this vertex-colours G in d colours.

8. $p_G(k) = p_{K_6}(k) + 4p_{K_5}(k) + 4p_{K_4}(k) + p_{K_3}(k)$
   $= k(k-1)(k-2)(k^3 - 8k^2 + 23k - 23).$

9. $p_G(k) = \dfrac{(k(k-1)(k-2)(k-3)(k-4))^2}{k(k-1)} + \dfrac{(k(k-1)(k-2)(k-3))^2}{k}$
   $= k(k-1)(k-2)^2(k-3)^2(k^2 - 7k + 15).$

11. (ii) With no restriction the number of ways is $k(k-1)^n(k-2)^{2n}$; with all the v_is the same the number of ways is $k(k-1)^n(k-2)^n$.

13. (iii) It is easy to check that any two non-isomorphic graphs on three or fewer vertices have different chromatic polynomials. So we need four vertices and, by (ii),

two non-isomorphic trees on four vertices will have the same chromatic polynomial; i.e.

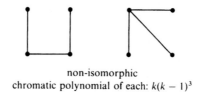

non-isomorphic
chromatic polynomial of each: $k(k-1)^3$

Chapter 12

1. No! For any rook polynomial the constant term is 1 and for any chromatic polynomial the constant term is 0.

2. (i) $1 + 12x + 48x^2 + 74x^3 + 40x^4 + 4x^5$.
   (ii) Number of ways with no i permed as stated $= 4$
   Number of ways with all i permed as stated
   $= 5! - 4! \times 12 + 3! \times 48 - 2! \times 74 + 1! \times 40 - 0! \times 4 = 8$.

3. The *shaded* board (in the hints) has rook polynomial
   $$(1 + 3x + x^2)(1 + 5x + 6x^2 + x^3) = 1 + 8x + 22x^2 + 24x^3 + 9x^4 + x^5.$$
   So the required answer $= 5! - 4! \times 8 + 3! \times 22 - 2! \times 24 + 1! \times 9 - 0! \times 1 = 20$.

4. The shaded board has rook polynomial
   $$(1 + 4x + 2x^2)(1 + 6x + 9x^2 + 2x^3) = 1 + 10x + 35x^2 + 50x^3 + 26x^4 + 4x^5.$$
   So the required answer $= 5! - 4! \times 10 + 3! \times 35 - 2! \times 50 + 1! \times 26 - 0! \times 4 = 12$.

5. (i) $\left(\left(\frac{n}{2}\right)!\right)^2$ if n is even and $\left(\frac{n-1}{2}\right)!\left(\frac{n+1}{2}\right)!$ if n is odd.

   (ii) $\left(\left(\frac{n}{2}\right)!\right)^2$ if n is even and 0 if n is odd.

9. (i) $1 + 1!\binom{n}{1}^2 x + 2!\binom{n}{2}^2 x^2 + 3!\binom{n}{3}^2 x^3 + \cdots + n!\binom{n}{n}^2 x^n$ (or equivalent).

   (ii) The *shaded* board has rook polynomial
   $$(1 + 4x + 2x^2)^4 = 1 + 16x + 104x^2 + 352x^3 + 664x^4 + 704x^5 + 416x^6 + 128x^7 + 16x^8$$
   and so the required answer is 4752, namely
   $$8! - 7! \times 16 + 6! \times 104 - 5! \times 352 + 4! \times 664 - 3! \times 704 + 2! \times 416 - 1! \times 128 + 0! \times 16.$$

(iii) $4752/8! = 33/280$.

(iv) This time the shaded board (shown in the hints) has rook polynomial

$$(1 + 16x + 72x^2 + 96x^3 + 24x^4)^{13}$$

and so calculation of $52! - 51! \times r_1 + 50! \times r_2 - \cdots$ certainly needs a computer. The probability of no 'snap' ($=$ this grand total divided by $52!$) is actually

$$\frac{4\,610\,507\,544\,750\,288\,132\,457\,667\,562\,311\,567\,997\,623\,087\,869}{284\,025\,438\,982\,318\,025\,793\,544\,200\,005\,777\,916\,187\,500\,000\,000} \quad (\approx 0.0162).$$

10. (i) $1 + \binom{m}{1}x + \binom{m-1}{2}x^2 + \cdots + \binom{m+1-k}{k}x^k + \cdots$

(ii) The rook polynomial of this board tidies up to

$$\sum_{k=0}^{n} \frac{2n}{2n-k} \binom{2n-k}{k} x^k$$

(iii) The number of ways of seating the $2n$ people is

$$2n! \sum_{k=0}^{n} (-1)^k \frac{2n}{2n-k} \binom{2n-k}{k} (n-k)!$$

(Note that our answer takes no account of the rotational symmetry of the table: it assumes quite naturally that each seat is different because, for example, the one nearest the door is very draughty!)

Chapter 13

2. (i) is planar and (ii) is not:

(i) 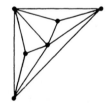 (ii)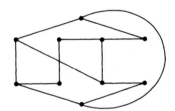

4. It is rather hard to give a convincing three-dimensional picture, so we use a standard topologist's method. Imagine each of the following figures on a piece of flexible and stretchable paper. Stick the side 'A1' of the paper onto side 'A2' to create a cylinder with the drawing on the outside. Then stick side 'B1' (which is now a circle) onto side 'B2' without any unnecessary twisting. The result will be K_7 (on the left) and $K_{4,4}$ (on the right) drawn without intersecting edges on the surface of a torus (like a doughnut with a hole in the

middle). This surface is geometrically/topologically equivalent to a plane with one extra handle.

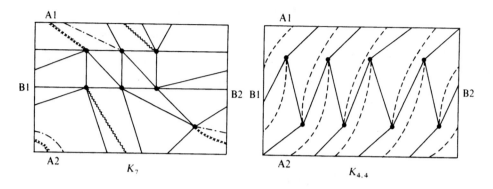

6. (i)

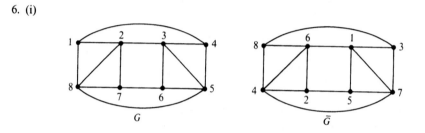

7. The number of regions inside the polygon is $\binom{n}{4} + \binom{n}{2} - n + 1$.

8. (i)

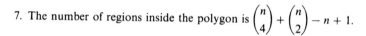

(ii)

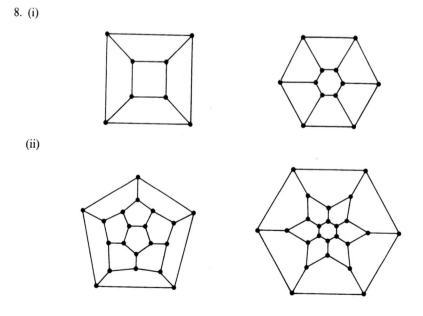

9. Yes, e.g.

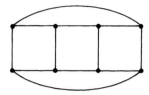

12. (c, d) can be $(3, 3)$, $(3, 4)$, $(4, 3)$, $(3, 5)$ or $(5, 3)$. A graph exists in each case (and in fact they are the 'maps' of the five famous 'Platonic solids' named):

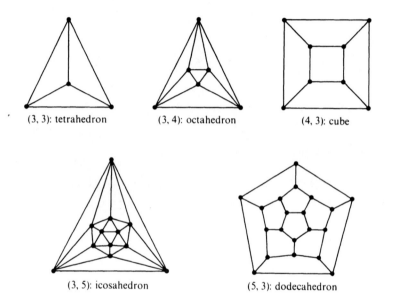

Chapter 14

2. If only three colours are allowed in the following arrangement then, starting at 1 and working upwards, all the numbered coins would have to be the same colour; this leads to two touching coins of the same colour:

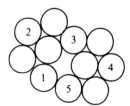

5. As in the answers to exercise 4 of the previous chapter, you can reconstruct a torus from the following picture by sticking edge A1 to A2 (without any unnecessary twisting) and then sticking B1 to B2. In the resulting 'map' on the surface there are seven faces and any two have a common boundary:

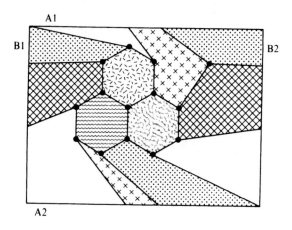

Chapter 15

7. One possible set of seven arrangements is

1 2 3	1 4 5	1 6 7	1 8 9	1 10 11	1 12 13	1 14 15
4 8 12	2 9 11	2 12 14	2 13 15	2 4 6	2 5 7	2 8 10
5 10 15	3 12 15	3 9 10	3 4 7	3 13 14	3 8 11	3 5 6
6 11 13	6 8 14	4 11 15	5 11 14	5 9 12	4 10 14	4 9 13
7 9 14	7 10 13	5 8 13	6 10 12	7 8 15	6 9 15	7 11 12

(This is, in fact, a special sort of Steiner triple system – as in exercise 8 – with $b = 35$ and $v = 15$.)

10. (iii) $L = \begin{pmatrix} \boxed{K} & & & & \\ & \boxed{K} & & & 0 \\ & & \boxed{K} & & \\ & & & \ddots & \\ 0 & & & & \\ & & & & \boxed{K} \end{pmatrix}$

11. (i) $p^2(3 - 2p)$ (ii) $p^4(15 - 24p + 10p^2)$

$p = 0.75 \Rightarrow$ (i) 0.844   (ii) 0.831

$p = 0.3 \Rightarrow$ (i) 0.216   (ii) 0.070

$p = 0.1 \Rightarrow$ (i) 0.028   (ii) 0.001

Chapter 16

11. As on page 205 arrange the $(r-1)(g-1)$ vertices in an $(r-1) \times (g-1)$ array and colour green just those edges joining two vertices in the same row. As we saw before there is no connected green graph on g vertices. Also, given any red graph, colour its vertices in row i in colour i ($1 \leq i \leq r-1$). This will give a vertex-colouring of any red graph in $r-1$ colours. Hence G_1 cannot be there in red.

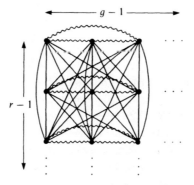

12. Let $g - 2 = k(r - 1)$. Then $r + g - 3 = (r - 1)(k + 1)$ and the red edges in K_{r+g-3} can form $k + 1$ components each consisting of a K_{r-1}, the remaining edges being green. Then there is clearly no red T_1. Also every vertex has just $g - 2$ green edges ending there. Therefore, as T_2 requires a vertex of degree $g - 1$, it follows that there is no green T_2.

Bibliography

M. Aigner: *Combinatorial theory* (Springer-Verlag, 1979)

I. Anderson: *A first course in combinatorial mathematics* (Oxford, 1979)

M. Behzad and G. Chartrand: *Introduction to the theory of graphs* (Allyn and Bacon, 1971)

L.W. Beineke and R.J. Wilson (ed): *Selected topics in graph theory* (Academic Press, 1978)

N.L. Biggs, E.K. Lloyd and R.J. Wilson: *Graph theory 1736–1936* (Oxford, 1976)

J.A. Bondy and U.S.R. Murty: *Graph theory with applications* (Elsevier, 1976)

J.A. Bondy and U.S.R. Murty (ed): *Graph theory and related topics* (Academic Press, 1979)

V.W. Bryant: *Yet another introduction to analysis* (Cambridge, 1990)

V.W. Bryant and H. Perfect: *Independence theory in combinatorics* (Chapman and Hall, 1980)

P.J. Cameron and J.H. Van Lint: *Graph theory, coding theory and block designs* (Cambridge, 1975)

S. Fiorini and R.J. Wilson: *Edge-colourings of graphs* (Pitman, 1977)

R.L. Graham, B.L. Rothschild and J.H. Spencer: *Ramsey theory* (Wiley, 1980)

F. Harary: *Graph theory* (Addison-Wesley, 1969)

R. Hill: *A first course in coding theory* (Oxford, 1986)

L. Lovasz and M.D. Plummer: *Matching theory* (North-Holland, 1991)

L. Mirsky: *Transversal theory* (Academic Press, 1971)

J.W. Moon: *Topics on tournaments* (Holt, Rinehart and Winston, 1968)

P.F. Reichmeider: *The equivalence of some combinatorial matching theorems* (Polygonal, 1984)

H.J. Ryser: *Combinatorial mathematics* (Maths Association of America/Wiley, 1963)

R.P. Stanley: *Enumerative combinatorics (vol. 1)* (Wadsworth and Brooks, 1986)

A.P. Street and D.J. Street: *Combinatorics of experimental design* (Oxford, 1987)

R.J. Wilson: *Introduction to graph theory (3rd ed)* (Longman, 1985)

Index

acyclic orientation 96, 137
alkane 24
alkene 24
Appel, K. 175
auxiliary equation 112

balanced incomplete block design 179–80
BIBD 179–80
binary code 192
binomial coefficient 1–12
binomial expansion 4
bipartite graph 30
block 179
board 140
boundary 156
bridge
  handle 161
  isthmus 75, 167
Brooks' theorem 129

capacity
  of a cut 105
  of an edge 103
castle 138
Catalan number 125, 250
Cayley's theorem 18, 111
chess-board 40, 140
Chinese ring puzzle 122
chromatic polynomial 133
circuit 14
class 1 (and 2) 89
closed path 72
code 192–6
coins 118, 178
colouring
  of edges 81–90, 172
  of faces 166–78
  of maps 166–78
  of vertices 127–37, 164, 172
complement
  of a board 144–5
  of a graph 79, 135, 163, 178

complete bipartite graph 22
complete graph 22
component 14
connected 14
convex hull 218
convex polygon 164, 218
convex polyhedron 154
convex set 218
coset 35, 213
cube 154, 259
cubic map 169
cut 105
cycle 14
cyclic design 184–8

δv 15
decreasing subsequence 38–9
degree 15
derangement 47, 146
design 179–92
dice 119–26
digraph 92
directed Eulerian path 93
directed graph 92
directed Hamiltonian cycle 93, 96
directed path 92
disconnected graph 14
disconnecting edge 75
disjoint paths 101
distance 128
distinct representatives 28
dodecahedron 154, 259
dominoes 40, 122
doubly stochastic matrix 34, 50
dual design 188–90
dual graph 176
duality of points and lines 66, 191

E 13
edge 13

edge-chromatic number 81
  of a bipartite graph 84
  of K_n 82
edge-colouring 81–90, 172
edge-disjoint 108
edge-set E 13
endpoint 13
error-correcting code 192–6
Euler, L. 71
Eulerian graph 72
Eulerian path 72, 93
Euler's formula 154–9, 162
Euler's function 46
Euler's square 60
Euler's theorem 71–4

face 152–4
face-colouring 166–78
factorial 3
family 28
Fáry's theorem 153
Fermat's last theorem 213
Fibonacci numbers 109, 113, 121
field 63, 187
file 48, 138
finite face 153
finite projective plane 65–9, 191
  and Latin squares 66–9
  of order n 65, 191
  of order 6 69, 192, 198
  of order 10 69, 192
Fisher's inequality 183
five-colour theorem 176
Fleury's algorithm 75
flow 103–8
flow-augmenting path 104–8
Ford, L.R. 105
four-colour theorem 168, 175
Fulkerson, D.R. 105

$\gamma(G)$ 161
$G = (V, E)$ 13
Galois field 65, 188
generating function 114–26
genus 161
  of $K_{m,n}$ 162
  of K_n 162
GF(n) 65, 188
girth 158
Graeco-Latin square 60
graph 13
grid 5

Hadamard matrix 199
Haken, W. 175

Hall's theorem 25–35, 108
  graph form 30
  harem form 91
  Latin squares 52–9
  marriage form 27
  matrix form 31, 98
  transversal form 29
Hamiltonian cycle 48, 77, 93, 96
Hamiltonian graph 77
handle 161
handshaking lemma 23, 226
harem problem 91
Heawood map-colouring theorem 176
homeomorphic graphs 159
hostess problem 150
Hungarian algorithm 101

icosahedron 259
icosian game 77
incidence matrix (of a design) 181
inclusion/exclusion principle 43–7, 145
increasing subsequence 38–9
in-degree 92–3
infinite face 153
injection 49
integer part 45
inversion 123
isomer 24
isomorphic graphs 22
isthmus 75

join 13

K-graph 159–60
$K_{m,n}$ 22
K_n 22
Kirkman's schoolgirl problem 197
Klein group 50
knight's moves 48, 77
König's theorem 84
König–Egerváry theorem 99, 108
Königsberg bridge problem 71
Kuratowski's theorem 160

Landau's theorem 95
Latin rectangles 50, 139
Latin squares 50–70
  and finite projective planes 66–9
  mutually orthogonal 60
  orthogonal 59–69
length (of a code) 192
longest path 15

Index

map   167
map-colouring   166–78
marriage theorem   25–35
matching   100
   from V_1 to V_2 30
max flow–min cut theorem   105
Menger's theorem   102
   edge form   108
minimax theorems   98–108, 195
mod n 63
modular arithmetic   63, 185, 187
multinomial coefficient   7–9
mutually orthogonal   60

$\binom{n}{k}$ 1

network   103–8

octahedron   259
officer problem   59–60
optimal assignment problem   101
order (of a projective plane)   65
orthogonal Latin squares   59–69
out-degree   92–3

paraffin   24
parameters of a design   180
parity   39–43
partition
   of integers   120–1
   of sets   116
Pascal's triangle   6, 109
path   14
perfect difference set   185–8
permutation   47, 139
permutation matrix   34
Petersen graph   158
pigeon-hole principle   36–39
planar graph   151–65
planar representation   152
Platonic solids   259
(la) problème des ménages   150
propane   250
Prüfer code   19

quadratic residues   187

Ramsey number   201, 204–6
Ramsey's theorem   201–12
Ramsey theory   201–23
rank (chess)   48, 138

rank (regiment)   59
recurrence relation   109–26
reflected routes   226
regiments and ranks   59–60
representative   28
Ringel, G.   162, 176
rook   138
rook polynomial   138–50
round-robin tournament   92–6
route in a grid   5–7
   reflected   226
   upper route   125

Schur's theorem   213
score   94
semi-Eulerian graph   73
semi-Hamiltonian graph   80
separating set of edges
   in a graph   108
   in a network   105
separating set of vertices   101
shortest path   128
sieve of Eratosthenes   49
simple graph   13
sink   105
snap   149–50
source   105
sprouts   165
Stanley's theorem   137
star-shaped set   153
Steiner triple system   197
Stirling numbers
   of first kind   117, 125–6
   of second kind   117, 125–6, 149
straight-line representation   153
strongly connected   93, 96
subdivision   159
subgraph   14
surjection   49, 126
symmetric design   184–92

t-design   180
tetrahedron   259
thickness   163
topology   152, 161
torus   163, 260
tournament   92–6
Tower of Brahma   110
Tower of Hanoi   110
trail   14
transversal   28
transversal theory   28, 31
tree   14–24, 197

unsaturated   104
upper route   125

V 13
(v, b, r, k, λ) design 180
value (of a flow) 105
van der Waerden's theorem 218
variety 179
vertex
   of a graph 13
   of a polygon 154, 218
vertex-chromatic number 127, 178
vertex-colouring 127–37, 164, 172
vertex-set V 13
Vizing's theorem 84, 88, 175

weight (of codeword) 195

Youngs, J. 162, 176

$\mathfrak{A}$ 28
$\gamma(G)$ 161
δv 15
$\binom{n}{k}$ 1
$n!$ 3
$[x]$ 45
$\{x\}$ 162
$|X|$ 23
$\chi(G)$ 126
$\chi_e(G)$ 81

CPSIA information can be obtained at www.ICGtesting.com
Printed in the USA
BVOW041915280812

299009BV00003B/27/A